# PREPARING OUTER ISLANDS FOR SUSTAINABLE ENERGY DEVELOPMENT

## MALDIVES' EXPERIENCE IN DEPLOYING ADVANCED HYBRID RENEWABLE ENERGY SYSTEMS

AUGUST 2024

ASIAN DEVELOPMENT BANK

© 2024 Asian Development Bank
6 ADB Avenue, Mandaluyong City, 1550 Metro Manila, Philippines
Tel +63 2 8632 4444; Fax +63 2 8636 2444
www.adb.org

Some rights reserved. Published in 2024.

ISBN 978-92-9270-859-7 (print); 978-92-9270-860-3 (PDF); 978-92-9270-861-0 (ebook)
Publication Stock No. TCS240404-2
DOI: http://dx.doi.org/10.22617/TCS240404-2

Note:
In this publication, "$" refers to United States dollars and "Rf" refers to rufiyaa.

Cover design by Josef Ilumin.

# CONTENTS

# TABLES AND FIGURES

# FOREWORD

by Thoriq Ibrahim, Minister of Climate Change, Environment and Energy of the Republic of Maldives

Maldives is almost entirely dependent on imported fossil fuel as a primary source of energy used for electricity generation and transport. This large dependence raises the cost of energy and makes the country very vulnerable to any adverse impact on the availability and price of fossil fuel in the international market. The government spends about $150 million annually to subsidize imported fuels to make the cost of electricity more affordable to Maldivians. To reduce this dependency, increase energy security, and achieve the government's target to cover 33% of the nation's electricity needs with renewable energy within the next 5 years, the government is putting all efforts needed in developing renewable energy systems across the country.

The Preparing Outer Islands for Sustainable Energy Development (POISED) project, implemented with support from the Asian Development Bank, has become a cornerstone in designing Maldives' just energy transition. POISED contributes to the resiliency of the benefiting islands by improving energy and food security, diversifying the economy, and creating jobs. POISED finances the replacement of inefficient diesel-based power generation grids in 160 outer islands with renewable-energy-ready grid systems that combine solar photovoltaic panels, batteries, modern diesel gensets, energy management systems, and upgraded distribution grids. These hybrid power systems have resulted in an average fuel savings of 25% per system. We are very grateful that the POISED project has also decisively supported significant cross-sector activities during the last decade such as the largest gender-sensitive training and capacity-building program in renewable energy covering 160 outer islands. The POISED project also supports the piloting and testing of a renewable energy-based ferry and solar photovoltaic community ice-making plants, creating jobs with a special focus on reducing gender inequality.

I would like to extend my gratitude to the Asian Development Bank, the Japan Fund for the Joint Crediting Mechanism, the Scaling Up Renewable Energy Program, and the European Investment Bank for the finance provided to implement the POISED project. I am also thankful to the State Electric Company Ltd., Fenaka Corporation Ltd., Island Councils, and the staff of the Energy Department of the Ministry for their support and commitment shown in the implementation of the project and to achieve the vision to provide sufficient, reliable, sustainable, secure, and affordable energy for a prosperous Maldives.

**Thoriq Ibrahim**
Minister of Climate Change, Environment and Energy
Republic of Maldives

# FOREWORD

by Takeo Konishi, Director General, South Asia Department, ADB

The Asian Development Bank (ADB) and the Government of Maldives have worked closely since 2014 to support the utilities in Maldives in transitioning from diesel to hybrid renewable energy mini-grid systems through the Preparing Outer Islands for Sustainable Energy Development (POISED) project. POISED is the largest energy sector intervention for Maldives with a target of 30.2 megawatt-peak solar photovoltaic installations, 12.5 megawatt-hour battery energy storage systems (BESS), and energy management systems (EMS) fully commissioned in 160 islands by the end of 2026. POISED also includes distribution network upgrades to allow high levels of renewable energy penetration.

Under POISED, the Government of Maldives and Fenaka Corporation Ltd. (FENAKA) decided to pilot test a technologically advanced hybrid system in Addu City, the second most populated island of Maldives. The pilot project was prepared and designed to apply for the Japan Fund for the Joint Crediting Mechanism (JFJCM). A $5 million grant was approved in 2015 to support the implementation of the hybrid system in Addu City, which is the first ever funding commitment made by the JFJCM. The hybrid system design in Addu City comprises 1.6 megawatt-peak of solar photovoltaic modules, a 0.5 megawatt-hour 3C rated lithium-ion BESS that can supply a maximum power output of 1.5 megawatts for 20 minutes, and an advanced EMS that ensures the readiness of BESS to meet any variation of the forecasted net demand.

The technologically advanced BESS and the EMS installed at Addu City using the JFJCM grant funding have improved the quality and reliability of electricity supply. Training of FENAKA staff was also provided as part of the pilot project to achieve sustainability and to influence similar systems implemented in the other islands. The JFJCM intervention in Addu City has increased the benefits of the POISED project, demonstrating how technological advancements can further improve renewable energy integration projects in small island nations. The capacities developed under this intervention, and in general under the POISED project, will enable FENAKA to take an active and leading role in catalyzing renewable energy development by the private sector.

ADB will keep supporting Maldives in its transition to renewable energy and in building resilience to the effects of climate change. ADB is preparing the Accelerating Sustainable System Development Using Renewable Energy project, the successor of POISED, to attain higher levels of renewable energy penetration through technological advances. More importantly, ADB is also deepening its engagement with project partners and civil society to mainstream gender equity and social inclusion.

**Takeo Konishi**
Director General
South Asia Department, ADB

# FOREWORD

by Bruno Carrasco, Director General, Climate Change and Sustainable Development Department, ADB

With Asia and the Pacific on the frontline of the growing climate crisis, the Asian Development Bank (ADB) is increasingly positioning itself as the climate bank for the region. ADB is reinforcing the climate and private capital mobilization objectives of its Strategy 2030, and implementing a Climate Change Action Plan that was released at the 28th Conference of the Parties to the United Nations Framework Convention on Climate Change in Dubai in 2023.

ADB will scale up its impact by catalyzing high-quality climate action through internal and external sources of finance and incentivize innovation in climate action with its own work and through partners to enhance its climate impact. ADB has an ambition to provide $100 billion in climate financing from its own resources from 2019 to 2030, of which at least $34 billion will be for climate adaptation and resilience. In 2023, ADB committed $9.8 billion in climate finance, which included $5.5 billion for mitigation and $4.3 billion for adaptation. In addition to our efforts, countries in the region need significant financing and support to meet their nationally determined contribution (NDC) targets and set higher ambitions in subsequent NDCs. This requires them to further mobilize both external climate finance and domestic resources for climate actions.

Maldives is one of the world's most vulnerable countries to the impacts of climate change. Despite its challenges and a negligible global share of greenhouse gas (GHG) emissions (only 0.003%), Maldives has been pursuing ambitious climate plans, aiming to reduce GHG emissions with financial support from the international community by 26% and to achieve net-zero carbon emissions by 2030 with extensive international assistance. Reducing the heavy reliance on oil imports, which helps power its 187 inhabited islands, will play an important role in not just achieving Maldives' NDC targets but also reducing its vulnerability to energy insecurity.

ADB mobilized co-financing from various internal and external sources, including the European Investment Bank, the Islamic Development Bank, the Japan Fund for the Joint Crediting Mechanism, and the Strategic Climate Fund, to support Maldives in implementing the Preparing Outer Islands for Sustainable Energy Development (POISED) Project. This project has been instrumental in helping Maldives transform its energy grids, reducing its dependence on imported diesel for electricity generation. Additionally, the project is contributing to alleviating the fossil fuel subsidy burden on the government budget, thereby reducing energy costs while decreasing GHG emissions.

This publication features the POISED project's contribution to improving energy systems in Maldives. It highlights the innovative component of the project in Addu City where advanced battery energy storage and energy management systems were introduced through carbon finance accessed through international carbon markets. This component is the very first project supported by the Japan Fund for the Joint Crediting Mechanism, which is a part of ADB's Carbon Market Program. Through this program, ADB has been providing carbon finance and technical and capacity-building support to developing member countries for climate mitigation initiatives. This project component is a good example to highlight the catalytic role carbon finance can play to incentivize investments in low-carbon technologies and solutions.

We hope the experience and lessons drawn from the POISED project will help other small island developing countries replicate the use of advanced low-carbon technologies to increase renewable energy supply, improve grid stability, and reduce energy cost, thereby contributing to achieving the targets of the Paris Agreement.

**Bruno Carrasco**
Director General
Climate Change and Sustainable Development Department, ADB

# ACKNOWLEDGMENTS

This publication on *Preparing Outer Islands for Sustainable Energy Development: Maldives' Experience in Deploying Advanced Hybrid Renewable Energy Systems* shares the experience of Maldives in transitioning toward increased use of renewable energy for power generation to reduce electricity cost and carbon dioxide emissions. The Preparing Outer Islands for Sustainable Energy Development project was supported by the Asian Development Bank, and this knowledge product was developed with the support of the Japan Fund for the Joint Crediting Mechanism (JFJCM).

Jaimes Kolantharaj (principal energy specialist, Energy Sector Group [SG-ENE]) conceptualized and supervised the development of this knowledge product, with the support of Shintaro Fujii (environment and climate change specialist and JFJCM fund manager, 2019–2022, Sustainable Development and Climate Change Department) and Tatsuya Yanase (environment and climate change specialist and JFJCM fund manager, Climate Change and Sustainable Development Department [CCSD]). The draft was prepared by Amila Wickramasinghe (energy economics expert), with technical inputs from Takahiro Murayama (climate change specialist, CCSD). Technical review was done by Priyantha Cabral Wijayatunga (senior director, SG-ENE) and Naresh Badhwar, Rastra Raj Bhandari, and Deborah Cornland (experts for the Asian Development Bank's Carbon Market Program); and Abdulla Firag and Sergio Ugarte (consultants, SG-ENE).

This knowledge product has also benefited from the technical review provided by Ahmed Ali of Maldives' Ministry of Climate Change, Environment, and Energy[1] and by Fenaka Corporation Ltd., the state-owned utility of Maldives.

Maria Charina Apolo Santos (project analyst, SG-ENE), Marietta Marasigan (consultant, SG-ENE), and Anna Blesilda Meneses (JFJCM fund coordinator [consultant], CCSD) provided coordination and administrative support.

---

[1]    Energy, environment, and climate change portfolios currently assigned to the Ministry of Climate Change, Environment, and Energy was previously held with the Ministry of Environment, Climate Change and Technology.

# ABBREVIATIONS

| | |
|---|---|
| ADB | Asian Development Bank |
| BESS | battery energy storage system |
| $CO_2$ | carbon dioxide |
| COVID-19 | coronavirus disease |
| DMC | developing member country |
| DSM | demand-side management |
| EIB | European Investment Bank |
| EIRR | economic internal rate of return |
| EMS | energy management system |
| FENAKA | Fenaka Corporation Ltd. |
| GHG | greenhouse gas |
| JCM | Joint Crediting Mechanism |
| JFJCM | Japan Fund for the Joint Crediting Mechanism |
| NDC | nationally determined contribution |
| O&M | operation and maintenance |
| POISED | Preparing Outer Islands for Sustainable Energy Development |
| PV | photovoltaic |
| SDG | Sustainable Development Goal |
| SFCR | specific fuel consumption rate |
| SIDS | small island developing states |
| STELCO | State Electric Company Ltd. |

# WEIGHTS AND MEASURES

| | |
|---|---|
| kWh | kilowatt-hour |
| MW | megawatt |
| MWh | megawatt-hour |
| MWp | megawatt-peak |

# CURRENCY EQUIVALENTS
(as of 31 May 2024)

Currency unit – Maldives rufiyaa (Rf)

Rf1.00 = $0.065

$1.00 = Rf15.42

# INTRODUCTION

## Background

Maldives is an archipelago in the Arabian Sea of the Indian Ocean comprising 1,212 islands. The islands are located within 20 atolls, forming a north-south-oriented ring spreading over an area of about 115,300 square kilometers. The total landmass of these islands is about 300 square kilometers, making Maldives the smallest country in South Asia. However, only 187 islands of Maldives are inhabited, with a population of about 540,000. The most populous island is Malé, the capital and the commercial hub of the country. One hundred and fifty-nine islands are designated as resort islands, while 30 to 40 are allocated for agriculture.[1] More than half of the country's population lives in the outer islands, engaging in fishing, agriculture, and tourism sectors. Tourism is bringing in more than 60% of the country's foreign income. Among the South Asia countries, Maldives has the highest per capita income, recording a per capita gross domestic product of $11,817.5 in 2022.

Except for a few islands where the electricity supply is provided by the island councils, the two state-owned electricity utilities—State Electric Company Ltd. (STELCO) and Fenaka Corporation Ltd. (FENAKA)—are responsible for supplying electricity to the inhabited islands. The island resorts, which are not serviced by the two electricity utilities, maintain their own captive generation. Despite the challenges posed by the physical separation of the land masses, in 2008, Maldives became the first South Asia country to achieve 100% electrification. This feat was achieved at the cost of having to develop and maintain isolated grids for each inhabited island.

The Maldives' isolated power grids are supplied using generators fired with imported diesel. This has made electricity supply expensive. The country's electricity tariffs, especially for businesses, have been among the highest in South Asia. Tariffs for business customers ranged from $0.20 to $0.40 per kilowatt-hour (kWh) in 2019.[2] The country's dependency on imported diesel also causes a significant reliance on foreign income earned through the tourism industry.[3] To reduce electricity prices, the government has offered fuel and electricity subsidies exceeding $65 million annually to some consumer categories (such as domestic consumers).[4] Maldives is also highly vulnerable to risks in security of supply and to fluctuations in international oil prices. A study by the United Nations Development Programme describes Maldives as the most vulnerable to oil price variations among the countries in Asia and the Pacific.[5] Diesel-based power generation is also a significant source of greenhouse gas (GHG) emissions.

---

[1]   Government of Maldives, Ministry of National Planning and Infrastructure, National Bureau of Statistics. 2021. *Statistical Yearbook of Maldives 2021*.

[2]   Government of Maldives, Utility Regulatory Authority. 2019. *Tariff Revision 2019*.

[3]   Even with the reduction of oil prices in the international market, Maldives spent $465 million on oil imports in 2019, owing to the increased demand for petroleum oil in the country.

[4]   Asian Development Bank (ADB). 2020. *A Brighter Future for Maldives Powered by Renewables: Road Map for the Energy Sector 2020–2030*.

[5]   United Nations Development Programme. 2007. *Overcoming Vulnerability to Rising Oil Prices: Options for Asia and the Pacific*.

# Maldives Energy Sector

Until 2009, STELCO was functioning as the vertically integrated utility responsible for electricity generation and delivery to consumers for about 40 islands across Maldives. Cooperative councils established by the island communities provided electricity for the other islands. From six islands that had reliable 24-hour electricity supply in 1990, Maldives succeeded in providing all inhabited islands with access to electricity around the clock by 2008. Upon achieving this, the government assigned six new regional utilities the responsibility for supplying electricity to the regions not served by STELCO. In 2012, these regional utilities were amalgamated to form FENAKA, the new utility responsible for providing electricity, water, and sewerage services to the outer islands.

Over the last 2 decades, Maldives has experienced a significant increase in energy consumption. From 2002 to 2011, energy demand increased from 224,000 to 396,000 tons of oil equivalent at a rate of about 6% per annum. The power sector was the key driver for this increase in energy demand. In 2018, Maldives electricity generation capacity was close to 530 megawatts (MW). Out of this installed generating capacity, 319 MW was to meet the electricity supply requirement in inhabited islands, while 210 MW was to supply resort islands.[6] Owing to the scattered geographic distribution of the population across 187 islands, with more than half of the population living in the outer islands, responding to increasing electricity demand was a challenge.

Maldives has significant renewable energy resources from solar and wind power. Multiple studies indicate that the use of renewable energy to complement diesel power generation in Maldives would reduce electricity production costs. Thus, a transition to renewable energy has a sound economic rationale. Accordingly, the government has taken various policy decisions supporting the increased use of renewable energy for electricity generation in the country.

The Government of Maldives has recognized the prospects of using the country's large renewable energy resources to lower the cost of electricity generation, increase energy security while reducing import-related financial risks, and reduce the country's GHG emissions. Accordingly, the government has embarked in two separate projects—Preparing Outer Islands for Sustainable Energy Development (POISED) project and Accelerating Sustainable Private Investments in Renewable Energy project—along with the Scaling Up Renewable Energy Programs in low-income countries, which is a targeted program of the Strategic Climate Fund to support energy efficiency and renewable energy investments in the country.

---

[6] Government of Maldives, Maldives Energy Authority. 2020. *Island Electricity Data Book 2019*.

# Policy Background and Other Commitments of Maldives

Since the Seventh National Development Plan published in 2007,[7] energy policies and strategies of Maldives have focused on shifting to renewable energy as an alternative to fossil fuel. For example, the National Energy Policy and Strategy 2010—which had been effective when the government-initiated POISED and Accelerating Sustainable Private Investments in Renewable Energy projects were launched—had the following policy targets:

    (i)    provide all citizens with access to affordable and reliable electricity supply,

    (ii)    achieve carbon neutrality in the energy sector by 2020,

    (iii)    promote energy conservation and energy efficiency,

    (iv)    increase national energy security,

    (v)    promote renewable energy technologies,

    (vi)    strengthen the management capacity of the energy sector,

    (vii)    adopt an appropriate pricing policy for the energy sector,

    (viii)    ensure customer protection, and

    (ix)    enhance the quality of energy services.[8]

The strategic action plan 2019–2023 outlined development targets and priorities of the government,[9] including those relevant to the energy sector (Table 1.0). The coronavirus disease (COVID-19) pandemic and the subsequent economic crisis prevented many of these targets to be achieved in the planned timeline. The government is working to update the strategic action plan for the next period with updated targets for the energy sector.

**Table 1.0:** Energy-Related Policies and Targets Under the Strategic Action Plan 2019–2023

| Policy | Target |
| --- | --- |
| Ensure access to affordable and reliable supply of electricity for all citizens. | • By 2023, implement electricity subsidies as targeted subsidies provided based on income levels of the consumers.<br>• By 2023, reduce distribution inefficiency by upgrading the distribution network and setting targets for the utilities to limit distribution losses to 7%. |
| Expand and develop the renewable energy sector. | • By 2023, increase the share of renewable energy in the national energy mix by 20% compared to 2018 levels.<br>• By 2023, at least 10 megawatts of solar photovoltaic is installed under net metering regulation. |

*continued on next page*

---

[7]    Government of Maldives, Ministry of Planning and National Development. 2007. *Seventh National Development Plan 2006–2010: Creating New Opportunities*.

[8]    Government of Maldives, Ministry of Housing and Environment. 2010. *Maldives National Energy Policy and Strategy*.

[9]    Government of Maldives. 2019. *Strategic Action Plan 2019–2023*.

**Table 1.0** *continued*

| Policy | Target |
|---|---|
| Increase national energy security through diversification of energy sources for energy production and expansion of energy storage. | • By 2023, reduce diesel consumption for electricity generation by 40 million liters.<br>• By 2023, increase renewable energy storage capacity to 30 megawatt-hour. |
| Strengthen the institutional and regulatory framework of the energy sector. | • By 2021, the Utility Regulatory Authority for integrated utility services, established in 2020, is functional and took over the regulating responsibilities previously assigned to the Maldives Energy Authority.<br>• By 2023, new public infrastructure projects shall have provisions to install renewable energy to meet their energy requirement.<br>• By 2023, energy data is up to date, reliable, and utilized for policymaking. |
| Promote energy conservation and efficiency. | • By 2023, green labeling is implemented for the energy sector.<br>• By 2022, provisions for green procurement under the Public Finance Act are implemented. |

Source: Government of Maldives. 2019. *Strategic Action Plan 2019–2023*.

The Maldives Energy Act of 2021,[10] established the framework to make energy services available to consumers at a reasonable price, promote renewable energy sources, and ensure that the energy sector promotes sustainable development that is environmentally friendly and adaptable to climate change. Also, in 2015, the government introduced a policy framework to guide climate-change-related activities in alignment with the national laws, development plans, strategies, and other policies.[11] The Maldives Climate Change Policy Framework identified the following principles that would govern the county's approach to addressing the climate change challenge:

- climate leadership,
- intergenerational equitability,
- mainstreaming of climate change,
- international commitments,
- multilateral partnerships,
- transfer of technology, and
- climate resiliency.

The national energy policy and the Maldives Climate Change Policy Framework were adopted at a time when many international initiatives were being undertaken to reduce GHG emissions and to transition toward sustainable and renewable energy sources. The Sustainable Development Goals (SDGs), established by the United Nations General Assembly in 2015, was one such initiative that identified the importance of affordable and sustainable clean energy (SDG 7) in the development process.

---

[10] Government of Maldives. 2021. *Maldives Energy Act, No. 18 of 2021*.

[11] Government of Maldives, Ministry of Environment and Energy. 2015. *Maldives Climate Change Policy Framework*.

Reducing carbon dioxide ($CO_2$) emissions from fuel combustion for electricity production is a key target identified for Maldives under SDG 7. In addition to the SDGs, under the United Nations Framework Convention on Climate Change (UNFCCC), countries had formed a common alliance to fight against climate change and were meeting regularly to discuss and agree on specific activities that each country could take to contribute to the common cause.

The Paris Agreement under the UNFCCC was reached in December 2015. The Paris Agreement emphasized the need for decarbonization to protect the planet and society from the impacts of global warming.[12] Keeping the global average temperature increase below 2°C (preferably at 1.5°C) compared with pre-industrial levels is envisaged.

Maldives was one of the first four countries—all small island developing states (SIDS)—to ratify the Paris Agreement, signifying that the countries most vulnerable to the impacts of climate change were taking the lead in the battle against global warming. In line with the Paris Agreement goals, Maldives pledged in its nationally determined contribution (NDC) to voluntarily reduce its GHG emissions by 10% from business-as-usual by 2030 and to enhance that target to 24% if provided international support.[13] The adoption of these targets reflected Maldives' commitment to embark in decarbonizing its energy sector.[14]

---

[12] Decarbonization is generally understood as the process of eliminating the combustion of fossil fuel or the emission of GHGs that result from the combustion of fossil fuel.

[13] The NDC is a document requested for each party to the Paris Agreement that outlines and communicates its post-2020 climate actions that it intends to achieve.

[14] Government of Maldives, Ministry of Environment. 2020. *Update of Nationally Determined Contributions of Maldives.*

# THE PROJECT: PREPARING OUTER ISLANDS FOR SUSTAINABLE ENERGY DEVELOPMENT

**2**

## Introduction to the Project

Mini-grids powered by diesel generators have been used to accelerate rural electrification in the inhabited islands in Maldives. However, diesel generators are expensive to operate and maintain. Hybridizing existing diesel-powered mini-grids with renewable energy sources offers a flexible and cost-effective means of transforming the energy landscape of the country to become more economically and environmentally sustainable.

The Preparing Outer Islands for Sustainable Energy Development (POISED) project was inaugurated on 19 January 2015. It is the largest renewable energy intervention ever undertaken in Maldives. The Asian Development Bank (ADB) coordinated and largely supported the project to reduce the country's dependence on fossil fuel for power generation in the outer islands. The project aimed to introduce sustainable energy in the outer islands, thereby helping to reduce the cost of electricity, minimize $CO_2$ emissions, achieve considerable fuel savings, and reduce the burden of electricity subsidies on the government budget. There were two specific outputs identified in the project design:

> **Output 1:** Renewable-energy-ready grid systems developed for outer islands and the greater Malé region, which include design and installation of equipment for solar–diesel hybrid grids on about 160 islands. Initially, five sample subprojects and a control center were to be developed.
>
> **Output 2:** Capacity of the Ministry of Environment and Energy, STELCO, and FENAKA enhanced to implement renewable energy grid interventions.

The project entailed converting existing diesel-based electricity grids in 160 inhabited outer islands across the 20 atolls into renewable diesel hybrid power systems. A minimum of 21 megawatt-peak (MWp) of solar photovoltaic (PV) capacity was installed. The expected electricity generation by the solar PV systems was 27,600 megawatt-hour (MWh) per annum, offsetting 19,623 tons of $CO_2$ emissions annually.[15]

The project is implemented in four phases. Phase 1 included installing solar PV and diesel hybrid smart grid systems in five islands representing a cross section of the entire power system of Maldives. The five islands were Addu, Buruni, Goidhoo, Kurendhoo, and Vilingili.

---

[15] Government of Maldives, Ministry of Environment. 2019. *A Shift Towards Clean Energy in Maldives*.

Three types of system architectures were identified for the hybrid systems implemented under the POISED project:

**Type A:** Power systems with moderate renewable energy penetration (less than 40% of peak load) and no energy storage.

**Type B:** Power systems with high renewable energy penetration (40% to 100% of peak load) where battery storage provides grid stabilization.

**Type C:** Power systems with very high renewable energy penetration (beyond 100% of peak load) where battery storage can allow the system to operate for 4 to 6 hours without diesel gensets.

Phase 2 of the project included investments in 17 additional outer islands, and installation of 8 MW of energy-efficient diesel generators in the capital city Malé. In addition, the distribution networks in these islands were upgraded to enable the integration of renewable energy sources in the electricity networks. Fifty outer islands are to be converted to hybrid power systems under Phase 3, while 44 islands will be converted under Phase 4.

# Project Compliance with Prevailing Policies and Strategies

The objectives of the POISED project, including improving the sustainability of the existing electricity grids, aligned well with the objectives of the national energy policy and climate change policy framework. Throughout its implementation period, the POISED project was well aligned with and highly relevant to these policies and the first NDC commitments and subsequent NDC updates made by Maldives.

ADB's Strategy 2020 and its midterm review identified the need to support ADB developing member countries (DMCs) in achieving sustainable energy growth. The POISED project was designed to reflect this long-term ADB strategy.[16] The project was also consistent with the interventions identified for Maldives by ADB through the country operations business plan for 2013–2015.[17]

To complement the POISED project, the government is facilitating private sector investments in solar photovoltaic systems on larger islands that were initially expected to be undertaken by STELCO. In addition, with support from ADB's Private Sector Operations Department and the Strategic Climate Fund, the government is promoting a low-carbon energy pathway for private tourist islands through policy measures and investment interventions.

The POISED project remained relevant through updates and revisions to policy guidelines: the Maldives Energy Policy and Strategies 2016 and the Maldives Strategic Action Plan 2019–2023, which further emphasized the need for Maldives to become independent of fossil fuel in meeting the country's energy needs. The action plan combined the specific policy targets identified in the energy policy and the

---

[16] ADB. 2008. *Strategy 2020: Working for an Asia and Pacific Free of Poverty.*

[17] ADB. 2012. *Country Operations Business Plan: Maldives, 2013–2015.*

commitments made by the government though its NDC to becoming a low-emission economy and to consolidating the policy directions of the energy sector. The latest NDC target of reducing GHG emissions by 26% compared with business-as-usual by 2030 and the vision to achieve net zero by 2030 further strengthen the project rationale (footnote 14).

# ADB's Commitment in the Project

In 2013, prior to commencing the POISED project, ADB provided technical assistance to conduct feasibility studies and to analyze and design hybrid power systems for Maldives. After assessing project feasibility, ADB provided project preparatory technical assistance to prepare the project for ADB financing.

In September 2014, ADB pledged $55 million of grant funding to implement the POISED project. The grants provided by ADB for the project included the following:

- Grant 0409-MLD: $38 million from the concessional ordinary capital resources and Asian Development Fund,
- Grant 0410-MLD: $12 million from the Strategic Climate Fund, and
- Grant 0429-MLD: $5 million from the Japan Fund for the Joint Crediting Mechanism (JFJCM).

Later, the POISED project received an additional $50 million of cofinancing from the European Investment Bank (EIB).

In addition to financial support, throughout project implementation, ADB actively engaged in coordinating the implementation of the project by providing resources for the government's project management unit, reviewing procurement documents, monitoring project progress, and providing guidance on technical and commercial issues (including those associated with cofinanciers such as the EIB). The government kept ADB fully informed and updated on project progress through regular communication and monitoring reports.

Concurrently, ADB maintained discussions with the government to further strengthen sector policies that aimed to achieve financial and institutional sustainability. ADB policy dialogue has supported the overall development of the Maldives energy sector, including the development of a medium-term road map and an investment plan. Through the dialogue, strengthening the capacity of government institutions (such as the Utility Regulatory Authority and the Ministry of Climate Change, Environment, and Energy) was identified as priority. An action plan was developed to achieve important milestones in this endeavor: the Utility Regulatory Authority was to develop and issue regulations for approving investment proposals in the energy sector, issue licenses to grid operators, and set new consumer tariffs, which were to be implemented by June 2015. A specific action that arose from the ADB policy dialogue with the government was the appointment of independent directors with technical, legal, and financial expertise, which had helped improve the governance of the two utilities, STELCO and FENAKA.

# Project Implementation Arrangements

FENAKA, the government-owned utility responsible for supplying electricity to the outer islands, implemented the POISED project through multiple engineering, procurement, and construction contracts. Maldives' Ministry of Finance was the project executing agency overseeing the project implementation and the channeling of funds between the financing agencies and contractors. A project management unit comprising officials from the Ministry of Climate Change, Environment, and Energy; FENAKA; and STELCO was established with ADB assistance to coordinate project implementation, including activities financed by cofinanciers, such as the EIB. Project implementation units were also established in FENAKA and STELCO to assist in preparing an overall implementation plan and to manage the contracts and annual budgets.

Phase 1 was completed in 2017. By the end of 2019, the POISED project had implemented hybrid power systems in 48 islands across 8 atolls. In 2021, the battery energy storage system (BESS) in Addu City was commissioned. By the end of 2023, a total of 60 hybrid power systems installed under the POISED project were in operation in Maldives.

Table 2.1 lists the islands and the installed solar PV, battery, and diesel generator capacities of the hybrid power systems that had been commissioned until July 2022.

**Table 2.1:** Hybrid Power Systems Established Under the Preparing Outer Islands for Sustainable Energy Development Project

| Atoll | Island | Solar PV (kWp) | Battery (kWh) | Diesel Generator (kW) | Number of Generators |
|---|---|---|---|---|---|
| Addu (or Seenu) | Addu City | 1,601 | 500 | 0 | 0 |
| Baa | Goidhoo | 203 | 84 | 160 | 2 |
| Gaafu Alif | Villingili | 304 | 84 | 1,300 | 2 |
| Haa Alif | Baarah | 162 | 70 | 150 | 2 |
| | Dhihdhoo | 246 | 150 | 1,300 | 2 |
| | Filladhoo | 140 | 280 | 70 | 1 |
| | Hoarafushi | 330 | 150 | 350 | 1 |
| | Ihavandhoo | 320 | 150 | 700 | 2 |
| | Kelaa | 200 | 100 | 0 | 0 |
| | Maarandhoo | 142 | 280 | 60 | 1 |
| | Molhadhoo | 60 | 120 | 50 | 1 |
| | Muraidhoo | 139 | 240 | 140 | 2 |
| | Thakandhoo | 91 | 180 | 130 | 2 |
| | Thuraakunu | 100 | 60 | 50 | 1 |
| | Uligan | 125 | 240 | 80 | 1 |
| | Utheemu | 150 | 280 | 100 | 1 |
| | Vashafaru | 120 | 240 | 140 | 2 |

*continued on next page*

**Table 2.1** *continued*

| Atoll | Island | Solar PV (kWp) | Battery (kWh) | Diesel Generator (kW) | Number of Generators |
|-------|--------|---------------:|--------------:|----------------------:|---------------------:|
| Lhaviyani | Kurendhoo | 107 | 42 | 254 | 2 |
| Noonu | Fohdhoo | 92 | 60 | 64 | 1 |
| | Henbadhoo | 92 | 80 | 348 | 3 |
| | Holhudhoo | 101 | 160 | 520 | 1 |
| | Kendhikulhudhoo | 134 | 110 | 520 | 1 |
| | Kudafari | 58 | 70 | 0 | 0 |
| | Landhoo | 124 | 110 | 108 | 1 |
| | Lhohi | 58 | 60 | 160 | 1 |
| | Maafaru | 159 | 140 | 388 | 2 |
| | Maalhendhoo | 79 | 70 | 200 | 1 |
| | Magoodhoo | 69 | 80 | 144 | 2 |
| | Manadhoo | 196 | 240 | 840 | 2 |
| | Miladhoo | 79 | 80 | 280 | 1 |
| | Velidhoo | 172 | 160 | 648 | 1 |
| Shaviyani | Bilehfahi | 140 | 100 | 64 | 1 |
| | Feevah | 66 | 80 | 0 | 0 |
| | Feydhoo | 98 | 100 | 200 | 1 |
| | Foakaidhoo | 131 | 100 | 520 | 1 |
| | Goidhoo | 58 | 70 | 80 | 1 |
| | Kanditheemu | 137 | 150 | 160 | 1 |
| | Komandoo | 105 | 120 | 160 | 1 |
| | Lhaimagu | 117 | 100 | 308 | 2 |
| | Maaungoodhoo | 98 | 100 | 80 | 1 |
| | Maroshi | 58 | 70 | 224 | 2 |
| | Milandhoo | 247 | 280 | 400 | 1 |
| | Narudhoo | 127 | 100 | 0 | 0 |
| | Noomaraa | 69 | 70 | 64 | 1 |
| Thaa | Buruni | 96 | 42 | 120 | 1 |
| | **Total** | **7,500** | **6,152** | **11,634** | **56** |

kW = kilowatt, kWh = kilowatt-hour, kWp = kilowatt-peak, PV = photovoltaic.

Note: Numbers may not sum precisely because of rounding.

Source: Government of Maldives, Ministry of Environment. 2019. *A Shift Towards Clean Energy in Maldives.*

During 2020–2021, the awarding of contracts and project implementation were delayed because of the COVID-19 pandemic. However, electricity grids for the remaining outer islands are still expected to be hybridized by the project. Table 2.2 provides the list of remaining hybrid systems planned to be implemented in Maldives.

**Table 2.2:** Hybrid Power Systems Remaining to be Installed

| Atoll | Solar PV (kWp) | Battery | | Generator (kW) |
| --- | --- | --- | --- | --- |
| | | (kW) | (kWh) | |
| Addu (or Seenu) | 1,750 | 3,220 | 1,110 | 0 |
| Alif Alif | 1,155 | 780 | 390 | 1,065 |
| Alif Dhaal | 743 | 485 | 244 | 485 |
| Baa | 2,425 | 2,405 | 1,645 | 3,350 |
| Dhaalu | 1,430 | 1,100 | 850 | 2,010 |
| Faafu | 850 | 840 | 720 | 360 |
| Gaafu Alif | 1,460 | 1,548 | 784 | 1,730 |
| Gaafu Dhaalu | 1,375 | 1,400 | 875 | 990 |
| Gnaviyani | 700 | 0 | 0 | 1,600 |
| Haa Alif | 2,326 | 2,090 | 2,540 | 3,288 |
| Haa Dhaalu | 2,270 | 1,495 | 1,495 | 1,610 |
| Kaafu | 624 | 274 | 136 | 1,100 |
| Laamu | 3,160 | 2,760 | 1,755 | 2,480 |
| Lhaviyani | 300 | 223 | 84 | 150 |
| Meemu | 1,570 | 1,570 | 1,670 | 910 |
| Noonu | 2,145 | 1,420 | 780 | 3,570 |
| Raa | 2,000 | 2,035 | 1,115 | 580 |
| Shaviyani | 2,160 | 1,440 | 855 | 1,600 |
| Thaa | 2,850 | 2,852 | 2,192 | 1,780 |
| Vaavu | 408 | 270 | 135 | 360 |
| **Total** | **31,701** | **28,207** | **19,375** | **29,018** |

kW = kilowatt, kWh = kilowatt-hour, kWp = kilowatt-peak, PV = photovoltaic.

Source: Government of Maldives, Ministry of Environment. 2019. *A Shift Towards Clean Energy in Maldives*.

# Japan Fund for the Joint Crediting Mechanism Support for the Project

The JFJCM is a single-donor trust fund established in June 2014 and administered by ADB with $136.03 million in funding provided by the Ministry of the Environment, Japan.[18] The fund aims to provide financial incentives for adopting advanced low-carbon technologies in ADB-financed projects.

---

[18] As of 31 March 2024.

The JFJCM provides grants and technical assistance to ADB projects that utilize the Joint Crediting Mechanism (JCM).[19] The JFJCM is an integral part of ADB's Carbon Market Program, which is ADB's flagship program established in 2005 to provide innovative carbon-market support to stimulate climate mitigation activities in ADB's DMCs.[20]

Projects cofinanced or administered by ADB are eligible for the JFJCM support to cover incremental costs associated with advanced low-carbon technologies employed in the project. Ongoing ADB-financed projects that meet the JFJCM eligibility criteria can also obtain the JFJCM support. The JFJCM can provide financial support of up to $10 million as grant funds for sovereign and nonsovereign projects.

Eligible recipients of the JFJCM support are as follows:

- For investment projects—borrowers in ADB DMCs that have signed a bilateral agreement with the Government of Japan for the JCM at the time of the JFJCM funding approval.
- For technical assistance projects—eligible governments (described above) and ADB DMCs that have initiated discussions with the Government of Japan on the signing of a bilateral agreement for the development of the JCM.

The criteria in selecting projects for the JFJCM financial support are as follows:

- use of advanced low-carbon technologies that are expected to result in clear and long-term GHG emission reductions;
- contribution to the host country's development goals, including providing environmental and social benefits;
- technical feasibility and financial and economic viability; and
- operational experience, track record, and institutional capacity of the project developer.

The first funding commitment made by the JFJCM was for the POISED project. POISED supported the piloting of a technologically advanced hybrid power system in Addu City, where a superior battery technology and a sophisticated control system were used to optimize solar PV and diesel power generation in a hybrid system. The pilot project was designed for the purpose of accessing the JFJCM funding. A $5 million JFJCM grant to support the project was approved in 2015.

By March 2024, the JFJCM had committed funds for eight ADB-financed investment projects in Bangladesh, Indonesia, Maldives, Mongolia, and Palau totaling $52.45 million.

---

[19] The JCM is a bilateral carbon market mechanism initiated by the Government of Japan, in which 29 countries (18 of them are ADB DMCs) are eligible to participate in issuing carbon credits for projects implemented within the countries (as of February 2024). The 18 eligible countries are Azerbaijan, Bangladesh, Cambodia, Georgia, Indonesia, Kazakhstan, the Kyrgyz Republic, the Lao People's Democratic Republic, Maldives, Mongolia, Myanmar, Palau, Papua New Guinea, the Philippines, Sri Lanka, Thailand, Uzbekistan, and Viet Nam. Effective 1 February 2021, ADB placed a temporary hold on sovereign project disbursements and new contracts in Myanmar.

[20] The Carbon Market Program also includes trust funds such as the Asia Pacific Carbon Fund, the Future Carbon Fund, and the Climate Action Catalyst Fund. It supports carbon-pricing initiatives and carbon market development in DMCs through the Technical Support Facility and the Article 6 Support Facility, which provide technical, capacity, and policy development support through in-country workshops, regional platforms, roundtables, and knowledge products.

# ADDU CITY
# HYBRID POWER SYSTEM

## 3

## Hybrid Power System Installed in Addu City by the Project

Under Phase 1 of the POISED project, a solar PV system with a total capacity of 1.6 MWp was installed in Hithadhoo, the main administrative district of Addu City. The solar PV complemented 14 diesel generator sets at the central power plant in Hithadhoo, with a total installed capacity of 15.25 MW. However, the operational diesel generator capacity was reduced to 6.75 MW because of de-rating and the inability to run the generator sets efficiently and effectively for longer periods.[21] Table 3.1 presents the details of the hybrid system in Addu City.

**Table 3.1: Summary Details of the Addu City Hybrid System**

| Parameter | Capacity |
|---|---|
| Total installed capacity of diesel generators (MW) | 15.25 |
| Available diesel generator capacity (MW) | 6.75 |
| Total installed capacity of solar PV (MWp) | 1.60 |
| Daily peak demand (MW) | 6.00 |

MW = megawatt, MWp = megawatt-peak, PV = photovoltaic.
Source: Fenaka Corporation Ltd. 2018. *Central Power Station, Addu City, Maldives – Power System Evaluation Report.*

Addu City has typical small-island grids with relatively small electricity demand. High penetration of solar power without battery or sufficient generator backup in such grids results in the system peak capacity varying significantly over time (both days and seasons). Consumer demand patterns also tend to fluctuate. In such contexts, maintaining grid stability is a challenge for the power system operator.

The installed 1.6 MWp solar PV capacity in Addu City, where the system peak demand is only 6 MW, can supply more than 20% of the system demand through solar PV, when the solar radiation is high. Such a high share of solar PV generation can make the grid unstable as a sudden reduction in solar radiation can cause a corresponding drop in solar PV generation, without providing enough time for the diesel generators to ramp up their generation. To avoid grid instability under such conditions, use of the full solar PV capacity needs to be restricted.

---

[21] The reduction of the capacity at which a diesel generator can safely operate for prolonged periods is referred to as de-rating of the generator. De-rating can happen for various technical reasons, including changes in ambient conditions and general wear and tear occurring with extended use.

# Upgrade to Addu City Hybrid Power System Using the Japan Fund for the Joint Crediting Mechanism Financing

The solar PV battery hybrid systems implemented under the POISED project use standardized designs, including BESS using lithium-ion batteries and standard battery-charging and control systems. However, for the Addu City hybrid system, a technologically advanced BESS, combined with an energy management system (EMS), was piloted with the view to promote future replication within Maldives and in other SIDS.

The upgraded hybrid system in Addu City comprised 1.6 MWp of solar PV modules, a 0.5 MWh 3C lithium-ion BESS, and an advanced EMS. It has the capability to

(i)    store excess solar PV power generated during daytime for nighttime use,

(ii)   flatten the load curve to compensate for fluctuations in consumer demand and solar PV supply, and

(iii)  optimize diesel generator operation.[22]

A BESS with a 3C maximum charging and discharging rate can respond faster than a BESS with a typical 0.5C or 1C rating. The hybrid system with a 3C BESS is capable of meeting sudden variations in demand and supply much more effectively than a 0.5C or 1C battery of similar capacity. This enables the use of more intermittent renewable energy resources in the system's generation mix.

An EMS is an autonomous control system that is primarily used to ensure the optimal scheduling of available distributed energy sources and storage systems in small power systems. The system includes the information and communication infrastructure to facilitate continuous control and monitoring of the power system. An EMS combined with an energy storage addresses power system optimization challenges, enabling the system operator to minimize grid operating costs by maximizing the use of lower-cost power generation units.

The ability of the operations staff to effectively use the advanced systems implemented in the project was considered a key factor influencing the long-term sustainability and replication potential of the project. Therefore, training of FENAKA staff was included as part of the pilot project. By participating in project implementation and taking part in on-the-job training conducted by the contractor, FENAKA staff has acquired the skills necessary to effectively operate the EMS and the BESS and carry out basic maintenance activities. The capacity developed through the project is expected to have a spillover effect, with similar systems being implemented in other islands by FENAKA as simple replications and adaptations.

---

[22]  The C rating of a battery represents the maximum rate at which the battery can be charged or discharged. The battery capacity is given as the power at which the battery must be charged (or discharged) to get it fully charged (or discharged) in 1 hour. However, whether the battery can be charged (or discharged) at that rate depends on the battery's C rating. 1C charging (or discharging) rate corresponds to the power that would fully charge (or discharge) the battery in 1 hour. But if the battery has a maximum charging (or discharging) rate of 0.5C, the maximum power it can charge (or discharge) is only half of the power denoted by the capacity rating. Therefore, a 0.5 MWh battery with a maximum C rating of 0.5 can only provide a maximum power output of 0.25 MW. With the rated capacity of the battery, it can supply this maximum power for 2 hours. Accordingly, the 3C rated battery used in Addu City can supply a maximum power output of 1.5 MW for 20 minutes.

**Advanced battery energy storage system installed in Addu City.** Technologically advanced battery energy storage system was installed in Addu City (photo by Takahiro Murayama/ADB).

**Capacity building on the operation of the advanced energy management system and the battery energy storage system.** FENAKA staff are being trained to operate the advanced energy management system and the battery energy storage system (photo by Maldives' Ministry of Climate Change, Environment, and Energy).

# Functionality of the Battery Energy Storage System and the Advanced Energy Management System in Addu City

The EMS installed in Addu City is superior in its functionality compared with the typical EMS used in most hybrid power systems. The Addu City EMS can forecast electricity demand and the likely electricity generation by the solar PV system (Figure 3.1). The solar PV electricity production forecasts are performed based on data from weather forecasts for the island, ambient conditions, and the production characteristics of the installed solar PV system. The EMS schedules the diesel generators to just meet the net system demand beyond forecast solar PV electricity production, thereby minimizing the fuel consumption of the diesel generators. At the same time, the EMS ensures the readiness of BESS to meet deviations from forecast demand.

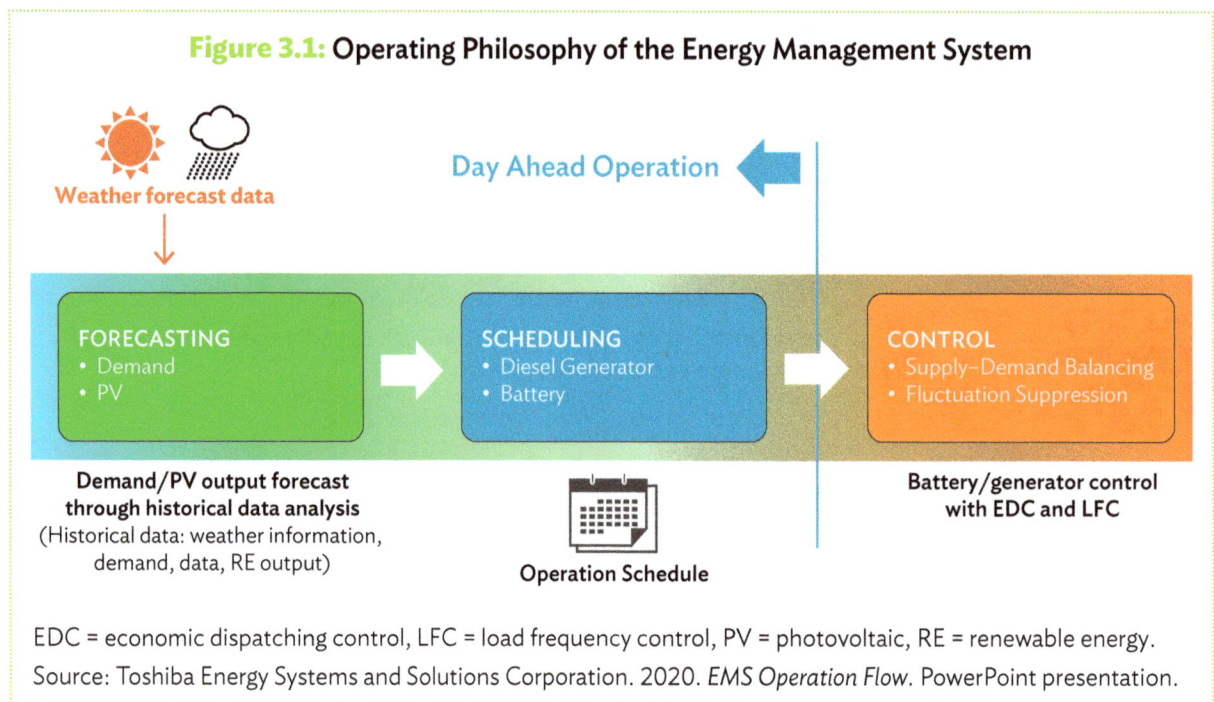

**Figure 3.1:** Operating Philosophy of the Energy Management System

Weather forecast data

Day Ahead Operation

**FORECASTING**
• Demand
• PV

**SCHEDULING**
• Diesel Generator
• Battery

**CONTROL**
• Supply–Demand Balancing
• Fluctuation Suppression

Demand/PV output forecast through historical data analysis
(Historical data: weather information, demand, data, RE output)

Operation Schedule

Battery/generator control with EDC and LFC

EDC = economic dispatching control, LFC = load frequency control, PV = photovoltaic, RE = renewable energy.

Source: Toshiba Energy Systems and Solutions Corporation. 2020. *EMS Operation Flow*. PowerPoint presentation.

Figure 3.2 shows how the EMS, with the help of the BESS, manages and controls the power system at near-perfect conditions, despite significant fluctuations in the solar PV generation. For example, on 3 August 2021, at 2:16 p.m., the solar PV generation dropped by about 40% in less than 30 seconds. Using the fast discharge capabilities of the BESS, the EMS ensured that the system frequency was maintained within 49.9 hertz and 50.0 hertz (i.e., a variation of only 0.2%).

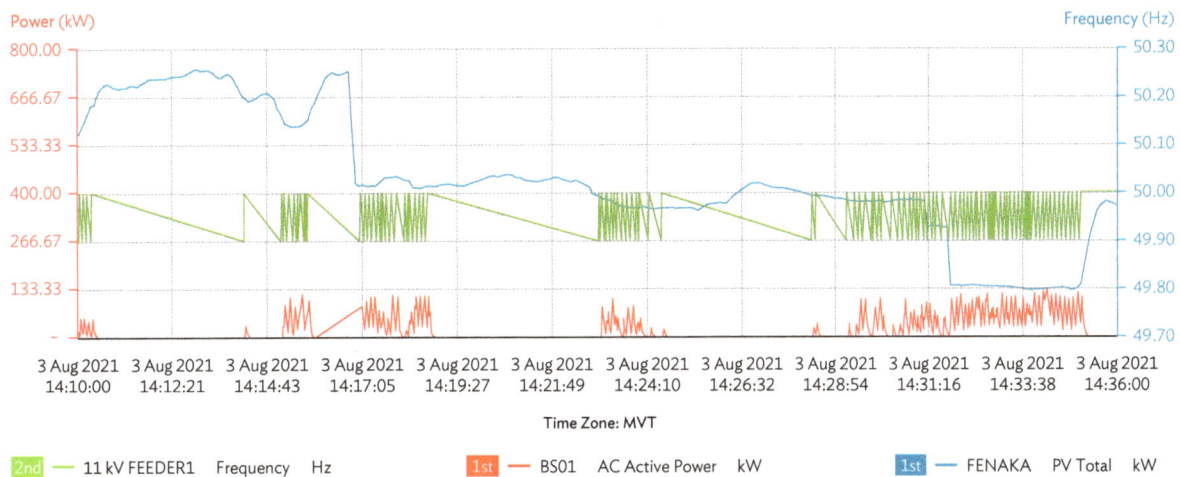

**Figure 3.2:** Net Demand Variations Managed by the Battery Energy Storage System and the Energy Management System

AC = alternating current, BS = battery storage, FENAKA = Fenaka Corporation Ltd., Hz = hertz, kW = kilowatt, MVT = Maldives time, PV = photovoltaic.

Source: Toshiba Energy Systems and Solutions Corporation.

When diesel mini-grids are hybridized by integrating solar PV, the average load on the diesel generators is reduced owing to the contribution of solar PV generation during daytime. Diesel generators tend to be less fuel efficient when lightly loaded; they can consume more fuel per kWh of energy produced (Figure 3.3). When this occurs, it reduces the emissions and cost benefits of solar PV generation.

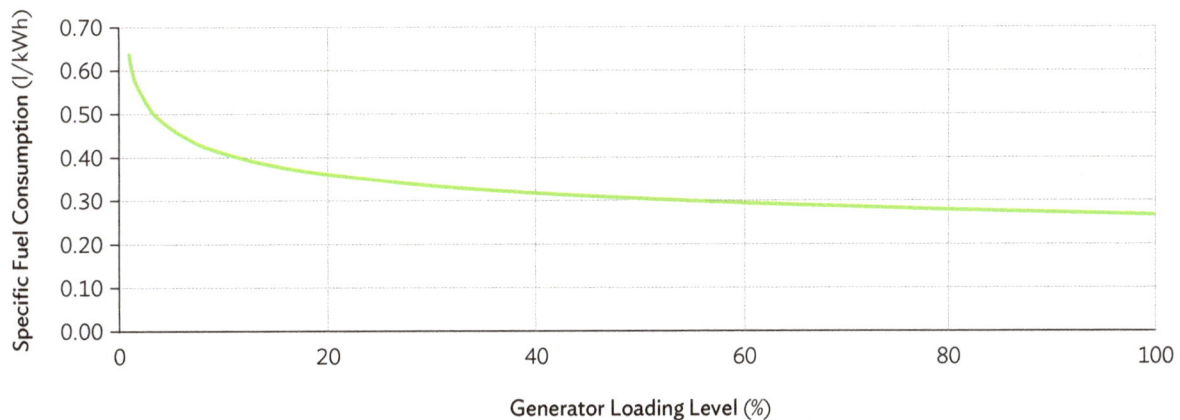

**Figure 3.3:** Variation of Specific Fuel Consumption Rate of Diesel Generators

kWh = kilowatt-hour, l = liter.

Source: J. Kersey, M. Sprengel, G. Babbitt, and T. Johnson. 2017. *Hybrid Power Generation for Improved Fuel Efficiency and Performance.* Conference Proceedings and Presentations of Electrical Energy Storage Applications and Technologies.

Figure 3.4 illustrates this effect by comparing the expected and actual reductions in fuel consumption in a typical diesel–solar hybrid system such as those used in the POISED project. The intermittent power supply of the solar PV results in the efficiency of the diesel generators dropping. In this example, 8.7% of the power requirement is met by solar PV. Despite offsetting that amount of diesel-based power generation, a corresponding reduction in fuel consumption is not achieved. It has been observed that only half of the expected fuel savings is achievable (Figure 3.4). This is because, when diesel generators are operated at varying output levels, more fuel is consumed to rapidly ramp up and down, compared with generators operating at a steady output level.

The Addu City pilot plant introduced a technologically advanced BESS, together with an EMS that has the capability to charge and discharge rapidly. The BESS addresses demand variations. When there is a sudden increase in demand, the BESS supplies that additional demand from its stored energy. When the power from generators or the solar PV exceeds demand, the BESS absorbs and stores the excess power. The EMS ensures that the diesel generators operate at their highest efficiencies, reducing overall fuel consumption. In this way, the need for generators to rapidly ramp up or down and waste fuel is also avoided. Therefore, the Addu City hybrid power system operates more efficiently than a typical hybrid system, resulting in lower operating costs.

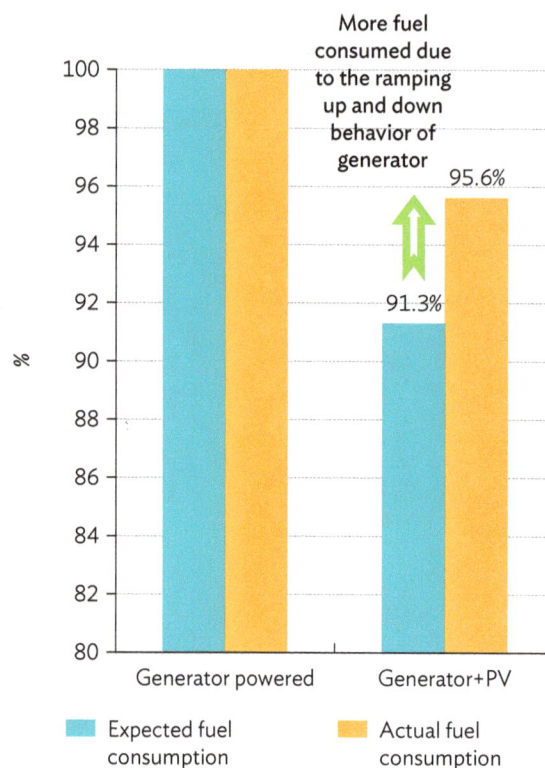

**Figure 3.4:** Example of Expected vs. Actual Fuel Consumption of a Diesel–Solar Hybrid System Installed by the Preparing Outer Islands for Sustainable Energy Development Project

PV = photovoltaic.
Source: Development Asia. 2019. *Advanced Battery Technology to Integrate Intermittent Renewables in the Maldives.*

Figure 3.5 illustrates how the actual fuel consumption of generators is reduced because of steady operation enabled by the advanced BESS and EMS. It has been estimated that the fuel consumption in Addu City has been reduced by 12.5%, exceeding the 8.7% reduction anticipated from the introduction of the solar PV generation by 3.8%.

Overall, the advanced BESS and the EMS installed in Addu City using the JFJCM grant funding has improved the quality and reliability of electricity supply while reducing the fuel consumed by diesel generators. The reduced fuel consumption results in both operating cost savings and GHG emission reductions. Table 3.2 presents the benefits gained from the POISED project and the JFJCM intervention.

**Figure 3.5:** Fuel-Efficiency Improvement from the Advanced Battery Energy Storage System and the Energy Management System

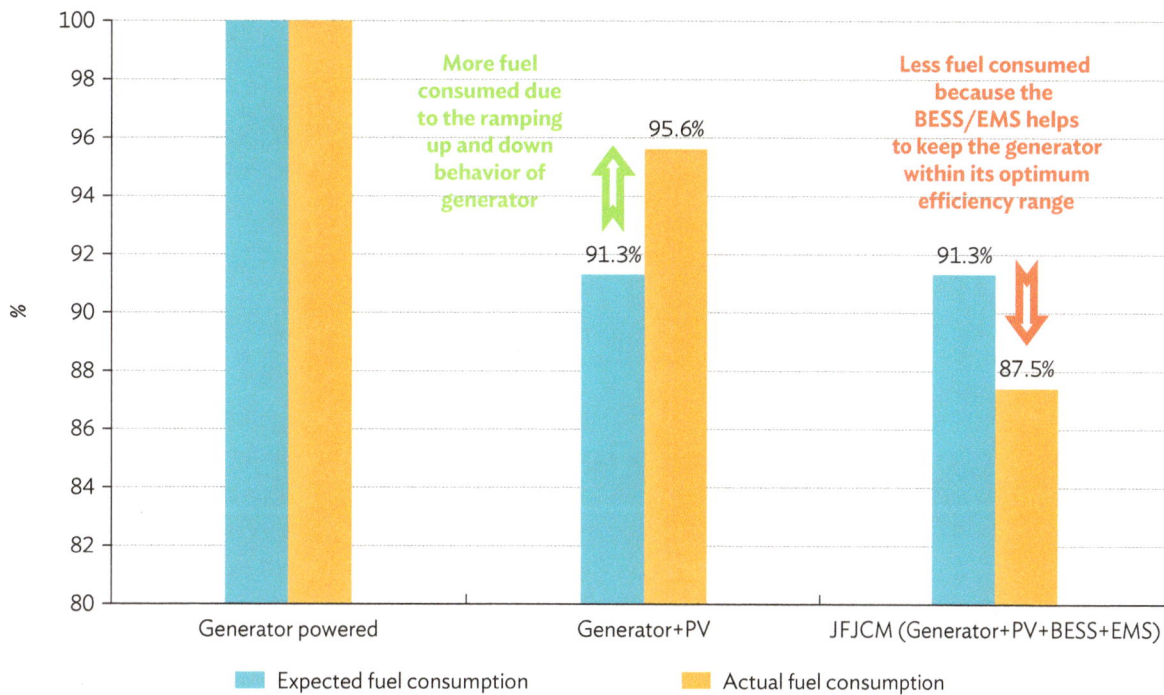

BESS = battery energy storage system, EMS = energy management system, JFJCM = Japan Fund for the Joint Crediting Mechanism, PV = photovoltaic.

Source: Development Asia. 2019. *Advanced Battery Technology to Integrate Intermittent Renewables in the Maldives.*

**Table 3.2:** Impact of the Preparing Outer Islands for Sustainable Energy Development Project and the Japan Fund for the Joint Crediting Mechanism on Operating Costs and Carbon Dioxide Emission

| Parameter | Generator Powered | Generator + PV | JFJCM (Generator + PV + BESS + EMS) |
|---|---|---|---|
| Diesel consumption rate (l/kWh) | 0.29 | 0.28 | 0.25 |
| Price of diesel (Rf/l) | | 8.00 | |
| Cost of diesel (Rf/kWh) | 2.32 | 2.22 | 2.03 |
| Cost of diesel ($/kWh) | 0.15 | 0.14 | 0.13 |
| $CO_2$ emission factor of diesel (kg$CO_2$/l) | | 2.68 | |
| $CO_2$ emission (kg$CO_2$/kWh) | 0.78 | 0.74 | 0.68 |

BESS = battery energy storage system, $CO_2$ = carbon dioxide, EMS = energy management system, JFJCM = Japan Fund for Joint Crediting Mechanism, kg = kilogram, kWh = kilowatt-hour, l = liter, PV = photovoltaic, Rf = rufiyaa.

Source: Asian Development Bank estimates.

# SUSTAINABILITY BENEFITS OF THE PREPARING OUTER ISLANDS FOR SUSTAINABLE ENERGY DEVELOPMENT PROJECT

**4**

## Environmental Benefits

The POISED project was initiated with the strategic agenda of contributing to inclusive and environmentally sustainable economic growth across Maldives by reducing the cost of electricity generation. The project also aimed to take urgent action to combat climate change, in line with SDG 13 (climate action), by reducing national GHG emissions. Transforming existing mini-grids in the outer islands into renewable diesel-based hybrid power systems contributes directly to reducing fossil-fuel consumption and associated GHG emissions. At the time when the project was designed, the power sector was the largest source of GHG emissions in Maldives, responsible for two-thirds of national GHG emissions.[23] Diesel is transported from the main harbors to each island using boats, which also consume fossil fuel, resulting in additional GHG emissions.

The POISED project also contributes to achieving SDG 6 (clean water and sanitation) by avoiding water contamination. Improved fuel-handling techniques are being used in the POISED project to prevent contamination of the country's limited freshwater resources. The project has the potential to increase its contribution to SDG 6 by providing low-cost electricity to water purification and desalination plants. Such plants could even be used to defer power load in future demand-side-management (DSM) activities implemented as part of ongoing and future projects.[24]

In the quest to build resilient infrastructure, which is part of SDG 9 (industry, innovation, and infrastructure), the POISED project included modifications to existing and new electricity supply infrastructure to climate-proof the project assets by making them resilient to rain and other climate-related hazards. Other infrastructure improvements made by the project that provide direct environmental benefits were establishing a proper method for removing and disposing leaked fuel and allocating storage capacity for used oil drums.[25]

---

[23] Government of Maldives, Ministry of Environment. 2019. *Maldives First Biennial Update Report to the United Nations Framework Convention on Climate Change.*

[24] A community outreach program to introduce household DSM in the outer islands is included in the POISED project, while further technical support on utility DSM interventions is envisaged in the future (ADB. 2014. *Report and Recommendation of the President to the Board of Directors: Proposed Grant and Administration of Grant to the Republic of Maldives for the Preparing Outer Islands for Sustainable Energy Development Project*). DSM is a strategy used by electricity utilities to control demand by encouraging consumers to modify their level and patterns of electricity usage.

[25] Government of Maldives, Ministry of Environment and Energy. 2014. *Proposed Grant and Administration of Grant—Republic of Maldives: Preparing Outer Islands for Sustainable Energy Development.*

# Social Benefits

The POISED project has also provided social benefits. Reducing air pollutants contributes to achieving SDG 3 (good health and well-being). Diesel power generators emit local air pollutants such as nitrogen oxides and particulate matter, exposing local communities to health hazards.[26] Reducing emissions of local pollutants from electricity generation by introducing solar PV reduces community exposure to such pollutants.

During project development, citizens of nearby local communities were able to generate income through employment in activities such as land clearing, tree trimming, and civil construction work. Operating the hybrid power systems provides employment opportunities for performing tasks such as periodical dusting and washing of PV panels, gardening, and minor maintenance work.

A gender focus was maintained during project design and implementation to ensure the participation of women in consultation processes and their access to employment opportunities during the construction period. The project included a gender action plan, where the project management unit of FENAKA was required to maintain gender equity throughout the implementation process, including training women to maintain the hybrid systems.

# Economic Benefits

Addition of electricity generated using solar PV to the grid by the POISED project leads to a corresponding reduction in electricity generation by diesel generators, substantially reducing the overall cost of grid electricity. The higher the contribution of solar PV, the lower the overall cost of electricity supplied to consumers. Thus, the replacement of fossil-fuel-based electricity generation with solar PV is expected to gradually reduce consumer tariffs, directly contributing to achieving SDG 7 (affordable and clean energy).[27] Furthermore, the diversification of energy resources used for electricity generation reduces the country's exposure to the impacts of volatile international crude oil prices.

Reduced consumption of imported diesel can provide indirect benefits to the economy by reducing the burden on public resources used to subsidize the fuel for electricity generation. Reducing the negative impacts of air pollutants on public health also leads to reduced healthcare costs, freeing up public resources.

Tourism is the highest contributor to the economy of Maldives, accounting for 19.9% of gross domestic product in 2017 (footnote 25). Tourism is also responsible for a significant portion of total energy consumption, for providing tourists with luxury services including transportation, heating and cooling, and desalination. Introducing renewables to the power generation infrastructure of the islands can contribute to establishing a green image, opening the ecofriendly tourism market for Maldives.

---

[26] İ. A. Reşitoğlu, et al. 2014. The Pollutant Emissions from Diesel-Engine Vehicles and Exhaust Aftertreatment Systems. *Clean Technologies and Environmental Policy*. 7. pp. 15–27. Long-term exposure to gaseous and particulate emissions can result in life-threatening conditions such as malfunction of organs, respiratory infections, asthma, and heart disease.

[27] World Bank. 2019. *Accelerating Renewable Energy Integration and Sustainable Energy*. Project Information Document.

The tourism and energy industries can complement each other through their contributions to the country's economic growth.

Because generating electricity using solar PV is cheaper than using fossil fuel, the higher the solar PV penetration level in the electricity supply system, the better the system operational economics is. Since the JFJCM intervention in Addu City had increased the contribution of solar PV to the generation mix, the economic benefits of the POISED project were enhanced. The JFJCM demonstrated how technological advancements can further improve economic returns and enable investments in intermittent sources of power generation in SIDS.

## Case Study 1: Economic Evaluation of the Solar Photovoltaic Installation at Kurendhoo Island

Phase 1 of the POISED project installed a hybrid diesel–solar power generation infrastructure in the grids of five islands. One of them, Kurendhoo, is located in the southwestern side of Lhaviyani Atoll, 129 kilometers from the national capital, Malé, and 16.4 kilometers from the atoll capital, Naifaru. Kurendhoo Island is oval shaped, with its length nearly twice its width, and has a total land area of about 21 hectares.

Because of the large potential for generating power using renewable resources in the island, a Type B hybrid power system equipped with battery storage was installed (see Chapter 2, Introduction to the Project), utilizing load shifting to enable a high level of penetration of solar PV into the power system.[28]

### Hybrid System Installed at Kurendhoo Island

Two diesel generators with capacities of 150 kilowatts (kW) and 100 kW were already operational and supplying electricity to Kurendhoo. The POISED project installed a solar PV system with a capacity of 107 kWp, a lithium-ion battery energy storage unit with a storage capacity of 42 kWh, and an additional diesel generation capacity of 250 kW.

### Electricity Generation of the Kurendhoo Island Hybrid System

Table 4.1 presents the actual diesel power generation and the solar PV generation recorded during the month of August 2017 (i.e., after the solar PV system became operational).

While the total electricity demand of the island does not change significantly, the solar PV generation varies from month to month, depending on the monthly variation of solar irradiance. Based on the solar resource data presented in Appendix 1, the average monthly solar power generation is estimated at 14,193 kWh.

---

[28]  Load shifting involves storing electricity when the power generation is higher than demand, with the intention of using the stored electricity to meet part of the electricity supply requirement when generation is lower than demand.

**Table 4.1:** Diesel and Solar Photovoltaic Electricity Generation in Kurendhoo Island, August 2017

| Source | Units Generated (kWh) |
|---|---|
| Diesel generator | 122,757 |
| Solar PV | 15,783 |
| **Total** | **138,540** |

kWh = kilowatt-hour, PV = photovoltaic.
Source: Fenaka Corporation Ltd.

## Savings from Reduced Diesel Consumption

The fuel savings achieved by installing the solar PV system and BESS in Kurendhoo are presented in Table 4.2. Based on actual operational data of the power plant recorded in the months of June and August 2017 (i.e., before and after the installation of the solar PV system and BESS, respectively), the average specific fuel consumption rate (SFCR) of the diesel power generators is 0.310 liter per kWh when the solar PV system and BESS are available, and 0.351 liter per kWh when these systems are not yet available. The assumed price of diesel is Rf8 per liter ($0.52). The average monthly savings through avoided diesel consumption during the first year of operation of the Kurendhoo Island hybrid system was Rf80,763 ($5,238).

**Table 4.2:** Comparison of Monthly Electricity Generation in Kurendhoo Island, With and Without the Preparing Outer Islands for Sustainable Energy Development Project

| Parameter | Without the Solar PV System and BESS | With the Solar PV System and BESS |
|---|---|---|
| Generation by diesel power plant (kWh/month) | 138,540[a] | 124,347[a] |
| Generation by solar PV (kWh/month) | – | 14,193[a] |
| Total electricity generation (kWh/month) | 138,540 | 138,540 |
| Specific fuel consumption rate (l/kWh) | 0.351 | 0.310 |
| Actual diesel consumption (l/month) | 48,628 | 38,532 |
| Diesel savings due to solar and BESS (l/month) | – | 10,095 |
| Avoided cost (through diesel savings) (Rf/month) | – | 80,763 |
| Avoided cost (through diesel savings) ($/month) | – | 5,238 |

BESS = battery energy storage system, kWh = kilowatt-hour, l = liter, PV = photovoltaic, Rf = rufiyaa.

[a] Fenaka Corporation Ltd.

Source: Asian Development Bank estimates.

## Savings from Reduction in Operating Period of Diesel Generators

Load shifting enabled by the solar PV system and BESS reduces the period of time when the diesel generators are in operation. The operation and maintenance (O&M) costs of diesel generators vary, depending on the number of hours the generators are operated. Therefore, reductions in generator runtime enabled by the solar PV system and BESS result in additional economic savings. These savings can be quantified by comparing the monthly operating period of all generators with and without the solar PV system and BESS. Table 4.3 presents the average monthly operating hours for the Kurendhoo diesel generators before and after the introduction of the solar PV system and BESS.

**Table 4.3:** Generator Runtime Reduction from the Solar and Battery Energy Storage Systems

| Parameter | Monthly Operating Period of the Kurendhoo Generators (Hours) | | | |
| --- | --- | --- | --- | --- |
| | DG 1 (250 kW) | DG 2 (150 kW) | DG 3 (100 kW) | Total |
| Without solar and BESS | 421 | 486 | 536 | 1,443 |
| With solar and BESS | 454 | 316 | 329 | 1,099 |
| Running hours | Increased by 33 | Reduced by 170 | Reduced by 207 | Reduced by 344 |

BESS = battery energy storage system, DG = diesel generator, kW = kilowatt.
Source: Fenaka Corporation Ltd.

With the contribution of solar PV generation, the total operating period of diesel generators decreased by 344 hours per month (i.e., by about 24%). The new solar PV system and BESS have also enabled the system operator to prioritize the use of the new and more energy-efficient generator (DG1), thereby improving the average SFCR of the power system (see Case Study 1, Savings from Reduced Diesel Consumption).

## Cost–Benefit Analysis of the Hybrid Power System Implemented in Kurendhoo Island

A cost–benefit analysis of the hybrid power system installed in Kurendhoo Island was carried out, comparing the project investments and operational costs against the cost savings achieved by the project. Only the diesel-consumption-related cost savings were included in the analysis, as the benefits arising from the reduced operating periods of the older generators have not been quantified. The data used in the cost–benefit analysis is presented in Table 4.4.

The economic internal rate of return (EIRR) of the project was also estimated using the annual net cost savings of the solar PV system and BESS over a period of 25 years. One percent of the project's capital cost of Rf3,902,237 is allocated annually for the O&M of the solar panels, including cleaning. It is also assumed that the solar inverter will need to be replaced once during the evaluation period, after 12 years. Ten percent of the capital is allocated for the inverter replacement.

**Table 4.4:** Data Used in the Cost–Benefit Analysis of the Solar and Battery Energy Storage Systems Installed in Kurendhoo Island

| Parameter | Value |
|---|---|
| Electricity produced by the diesel generators (kWh/year) | 1,492,164[a] |
| Estimated electricity produced by the solar PV system (kWh/year) | 170,316 |
| Total electricity generation (kWh/year) | 1,662,480[a] |
| Specific fuel consumption rate (l/kWh) without the project | 0.351 |
| Specific fuel consumption rate (l/kWh) with the project | 0.310 |
| Diesel consumption without the solar PV system and BESS (l/year) | 583,530 |
| Diesel consumption with the solar PV system and BESS (l/year) | 462,386 |
| Avoided diesel consumption resulting from the solar PV system and BESS (l/year) | 121,144 |
| Price of diesel (Rf/l) | 8 |
| Annual cost savings from avoided diesel consumption (Rf/year) | 969,152 |
| Total project investment (Rf) | 3,902,237 |
| Simple payback period (years) | 4.03 |

BESS = battery energy storage system, kWh = kilowatt-hour, l = liter, PV = photovoltaic, Rf = rufiyaa.

[a] Fenaka Corporation Ltd.

Source: Asian Development Bank estimates.

The cost savings achieved from the solar PV system and BESS during the first year of operation is about Rf969,152. However, because the performance of the solar panels degrades over time, the cost savings may decrease by 1% annually. Table A1.3 in Appendix 1 provides the marginal annual O&M costs, savings, and net economic benefit estimates from the cost–benefit analysis.

The project is estimated to yield an EIRR of 23% (Appendix 1). This EIRR is substantially higher than the benchmarked EIRR values used to evaluate capital investments in the country. Typically, an investment yielding an EIRR above 10% is considered acceptable in Maldives. Thus, the 23% EIRR indicates that the POISED project in Kurendhoo was a highly efficient investment.

## Case Study 2: Economic Benefits of Using a Fast-Responding Battery Energy Storage System and Advanced Energy Management System in Addu City

An advanced BESS containing lithium-ion 3C batteries and an advanced EMS was installed at Addu City to upgrade the existing Type A diesel–solar hybrid system (i.e., a system without storage) that was installed in 2017 as part of the first phase of the POISED project. The Addu City project was expected to be a high impact pilot owing to the relatively large population that it serves and the corresponding electricity demand, as well as the presence of important infrastructure such as an airport.

## Cost Savings Achieved Through the Upgrade of the Power System in Addu City

As discussed in Chapter 3's *Upgrade to Addu City Hybrid Power System*, a BESS system containing fast-responding 3C lithium-ion batteries can compensate for rapid variations in solar PV generation and consumer electricity demand, avoiding the use of diesel generators. This allows the diesel generators to be operated steadily at optimal efficiency levels. In addition to displacing diesel-based power generation, adding a BESS and EMS to the solar PV system results in reducing the SFCR of the generators from 0.290 to 0.256 liter per kWh, further increasing fuel savings. Table 4.5 presents a comparison of the power generation, fuel consumption, and cost savings of the power system in Addu City, with and without the project.

**Table 4.5: Economic Impact of Upgrading the Power System in Addu City**

| Parameter | Without BESS and EMS | With BESS and EMS |
|---|---|---|
| Electricity generation by diesel generators (kWh/year) | 36,211,229[a] | 35,017,611[a] |
| Estimated electricity generation by solar PV (kWh/year) | 1,193,618 | 2,387,236 |
| Total electricity generation (kWh/year) | 37,404,847[a] | 37,404,847 |
| Specific fuel consumption rate (l/kWh) | 0.290 | 0.256 |
| Annual diesel consumption (l/year) | 10,847,406 | 9,313,807 |
| Avoided diesel consumption (l/year) | | 1,533,599 |
| Economic cost savings (Rf/year at Rf8 per liter) | | 12,268,792 |
| Economic cost savings ($/year) | | 795,642 |

BESS = battery energy storage system, EMS = energy management system, kWh = kilowatt-hour, l = liter, PV = photovoltaic, Rf = rufiyaa.

[a] Fenaka Corporation Ltd.

Source: Asian Development Bank estimates.

Through the system upgrade during the first year of operation of the fast-responding batteries, the annual average diesel fuel saved is 1,533,599 liters and cost savings is Rf12,268,792 ($795,642). This benefit can be attributed to the increased solar PV penetration and the improved operating efficiency of diesel generators, which resulted from using the fast-responding batteries to manage the intermittency of solar PV power generation and variations in consumer demand.

## Economic Evaluation of the Addu City Hybrid Power System Upgrade

To assess the benefit of using fast-responding 3C batteries instead of batteries with a 1C charging rate, the EIRR of the two investment options were compared. To ensure that the evaluation captures all operational benefits associated with 3C batteries, the comparison was made between the 3C battery capacity of 500 kWh installed in Addu City against a 1C battery capacity of 1,500 kWh that would provide the performance and operational flexibility equivalent to the 3C battery.

The capital cost of the Addu City pilot project investment, comprising 500 kWh of 3C fast-responding lithium-ion batteries, was about Rf50 million ($3.2 million). The capital cost of a hybrid system with a BESS comprising 1,500 kWh of 1C batteries is about Rf55 million ($3.6 million). The estimated capital cost of using conventional batteries is slightly higher because of the larger capacity of batteries required compared to the 3C battery option. In addition, both options involve a considerable cost for network and infrastructure improvements.

The 3C batteries used in the Addu City pilot project are expected to have a lifetime of 15 years, compared with the 8-year expected lifetime of conventional lithium-ion batteries. The battery replacement costs were calculated by assuming Rf10,042.50 per kWh ($651 per kWh) for the 3C batteries and Rf3,862.50 per kWh ($250 per kWh) for the 1C batteries.[29] The avoided cost of diesel, costs associated with the solar panels, and inverter replacement costs were assumed to be the same for both battery options. Based on the avoided cost of diesel presented in Table 4.5, savings from installing solar PV in the first year of operation is about Rf12,268,792 ($795,641). A 1% annual reduction in fuel savings is expected because of the degradation of solar panels. Half a percent of the initial investment is allocated annually for the O&M of the solar PV system, including cleaning and dusting of solar panels. Assuming the inverters have a 13-year useful operational life, the replacement of the solar inverter is expected at the end of the 13th year. Accordingly, 10% of the initial investment was allocated for the inverter replacement cost.

The annual costs and benefits of both battery options over the 25-year evaluation period are presented in Appendix 2. The net benefit streams and the corresponding EIRR of each battery option is also presented. The EIRRs of the two options suggest that, despite the higher unit cost of 3C batteries, their faster response rate and longer life make them more economical than the conventional 1C lithium-ion batteries in the long term. The EIRR when conventional batteries are used is 20% compared with 23% when using fast-responding 3C batteries. Additional information on the cost–benefit analysis comparing 1C and 3C batteries is provided in Appendix 2.

In addition to economic benefits, the longer useful life of the batteries used in the Addu City project reduces the need to dispose of or recycle batteries over the project life. Considering the environmental concerns associated with battery disposal, the technologically advanced batteries used in the Addu City project are a more environmentally sustainable solution. Such features could make the fast-responding 3C batteries the preferred technology for electricity storage in hybrid power systems in the future.

---

[29] W. Cole, A.W. Frazier, and C. Augustine. 2021. *Cost Projections for Utility-Scale Battery Storage: 2021 Update.* National Renewable Energy Laboratory.

# CONCLUSION AND POLICY RECOMMENDATIONS

**5**

## Lessons Learned

The POISED project implemented in Maldives infused a significant volume of funding. It was the largest ever coordinated allocation of resources for renewable energy investments in the country, supporting a paradigm shift in the approach to power generation in the island nation. The project introduced solar PV with control and electricity storage capabilities to maximize the benefit of using renewable energy in lieu of fossil fuel. The successful implementation of the POISED project (60 islands), and subsequent expansion to 160 islands with the EIB financing, demonstrated that replacing diesel-based power generation with solar PV is technically feasible and can bring substantial environmental, social, and economic benefits. Thus, scaling up the use of hybrid solutions should be seriously explored particularly by SIDS, other isolated locations such as remote communities, as well as industries and mines using diesel-generator-based power systems.

The operational flexibility and high investment returns achieved in Addu City by using technologically advanced batteries and a state-of-the-art EMS indicate that the combination of improvements in battery technology and EMS programs can provide economic benefits in the longer term, despite the higher capital costs of such technologies.

Implementation of the POISED project was negatively affected by the COVID-19 pandemic, which posed unanticipated difficulties for all of the stakeholders involved. Some activities were substantially delayed because of travel restrictions and unavailability of equipment and materials during the pandemic. This highlights the risk Maldives is exposed to with respect to the availability of expertise, equipment, and materials required for development, and emphasizes the importance of increasing the domestic skilled labor force and maintaining sufficient equipment and material inventories to ensure the sustainability of hybrid power projects. At the same time, the pandemic underscored the value of the improved energy security achieved through the POISED project.

Implementation of the EIB component of the POISED project is supporting the island grid upgrades with energy storage and control systems, which will enable the deployment of additional intermittent renewable energy investments by the private sector. This can serve as a model for leveraging increased private sector investment in renewable energy in the country, through appropriate public sector investment.

# Policy Recommendations

The Maldives National Energy Policy and Strategy 2010 and the country's international commitments in its NDC contain a strong focus on moving toward carbon neutrality by adopting renewable energy technologies. The POISED project aligns well with those policies and the Maldives Strategic Action Plan 2019–2023, which aims to operationalize them. The success of the POISED project confirms the benefits and potential to achieve the goals of achieving carbon neutrality, affordability, and energy security that are embedded in the policies, strategies, and action plans adopted by Maldives.

However, the experience gained through the POISED project suggests that it could be prudent to revisit the timeframes for targets in those policies and plans. A reassessment of target timeframes should consider the risks associated with stagnation of economic activity, as was experienced for almost 2 years as a result of the COVID-19 pandemic, and the lessons learned through the POISED project regarding the time required to implement transformational investments at scale (including financing, procurement, construction, and commissioning).

The JFJCM intervention through the POISED project illustrated the potential to use advanced technologies to improve conventional power systems, delivering higher technical performance and better economic returns. Currently, Maldives' policies are silent on exploring advanced technology options, which could lead to missed opportunities for technological leapfrogging by adopting the latest and best-performing technologies. Thus, it would be beneficial to include the need to explore innovative opportunities; research; and the development of new technologies, processes, and business models as an integral part of the country's efforts to shift toward more extensive use of renewable energy resources.

While the project was designed to improve supply-side efficiency, other initiatives mentioned in national energy policies and strategies, such as the implementation of demand-side management (DSM) and energy efficiency improvements, have not been addressed sufficiently. Taking more actions in these areas is therefore recommended to achieve policy targets and fulfill the NDC commitments. Inclusion of DSM in future energy projects could be considered and provided as policy guidelines in parallel with the promotion of low-carbon electricity generation.

Scaling up the use of solar PV in Maldives for power generation is constrained by land scarcity. Therefore, options such as rooftop solar, floating solar, wind, and other renewable energy resources, including marine renewable energy, should be explored. At the same time, Maldives should monitor trends in the cost of such technologies as well as changes regarding technical and financial constraints that currently prevent them from being utilized in Maldives.

# Future Development of Renewable Energy in Maldives

The success of the POISED project, especially with respect to the economic benefits it provided, has paved the way for the commercial development of similar investments by the private sector under different business models. With the capacity developed during the implementation of the POISED project, FENAKA could take an active and leading role in catalyzing such investments. Concessional financing was made available for the POISED project. Other models, such as carbon finance, should be explored to identify options when concessional financing and grants are not available for promoting investments in renewable energy projects.

The fast-responding BESS and the advanced EMS installed in Addu City, with the support of the JFJCM, demonstrated a new dimension of the hybridization of diesel mini-grids. Technical limitations that previously prevented increasing the intermittent renewable energy generation in hybrid power systems were minimized, thereby reducing the long-term cost of electricity production. The POISED project in Addu City demonstrated that a power system can be operated safely and stably even with a relatively high penetration level of intermittent sources of electricity. Addu City's hybrid system can be replicated to fully reap the benefits of hybrid diesel–solar systems using advanced BESS and EMS technology. Increasing the solar PV contribution to the electricity supply systems of other SIDS and small isolated mini-grid systems in other contexts is therefore worth pursuing.

# APPENDIX 1

# SOLAR PHOTOVOLTAIC GENERATION IN KURENDHOO ISLAND

The energy produced by the solar photovoltaic (PV) system in Kurendhoo Island was estimated using the PV Watts Calculator software tool.[1] The estimation is based on the specifications presented in Table A1.1. The tilt angle of the solar PV array was assumed to be 8° considering a 3° to 10° tilt-angle range for Maldives. The inverter efficiency was assumed at 96.0%, and the system losses at 14.1%. Because the solar PV array rarely produces more than 80% to 90% of its rated power, solar PV systems are often designed such that the PV system power rating is greater than that of the inverter. Therefore, the direct current to alternating current ratio was assumed to be 1.2.

**Table A1.1:** Specifications of the Solar Photovoltaic System in Kurendhoo Island

| Parameter | Kurendhoo Island, Lhaviyani Atoll, Maldives |
|---|---|
| Latitude (degree North) | 5.2 |
| Longitude (degree East) | 73.3 |
| Elevation (meter) | 0 |
| DC system size (kilowatt) | 107 |
| Module type | Standard |
| Array type | Fixed (open rack) |
| Array tilt (degree) | 8 |
| Array Azimuth (degree) | 180 |
| System losses | 14.1 |
| Inverter efficiency | 96.0 |
| DC to AC ratio | 1.2 |

AC = alternating current, DC = direct current.
Source: Fenaka Corporation Ltd.

The electricity generation estimates for the 107 kilowatt-peak solar PV system in Kurendhoo Island are presented in Table A1.2. Figure A1.1 presents the average solar radiation, while Figure A1.2 shows the estimated monthly electricity generation of the 107 kilowatt-peak solar PV system. The annual average solar radiation of Kurendhoo Island is about 5.92 kilowatt-hour per square meter per day (kWh/m²/day), and typically varies between 5.16 kWh/m²/day and 6.83 kWh/m²/day. The maximum (16,152 kWh/month) solar power generation level occurs in March and the minimum (12,442 kWh/month) occurs in June.

---

[1]    The PV Watts Calculator developed by the National Renewable Energy Laboratory of the United States is a software tool used in the solar PV industry to estimate the energy production of grid-connected solar PV energy systems.

**Table A1.2:** Estimated Monthly Solar Power Generation from the 107-Kilowatt-Peak System in Kurendhoo Island

| Month | Solar Radiation (kWh/m²/day) | Solar PV System Output (kWh) |
|---|---|---|
| January | 6.27 | 15,439 |
| February | 6.83 | 14,840 |
| March | 6.79 | 16,152 |
| April | 6.18 | 14,336 |
| May | 5.47 | 13,598 |
| June | 5.16 | 12,442 |
| July | 5.28 | 13,073 |
| August | 5.78 | 14,273 |
| September | 5.90 | 13,960 |
| October | 6.19 | 15,233 |
| November | 5.60 | 13,140 |
| December | 5.61 | 13,838 |
| **Average** | **5.92** | **14,193** |

kWh = kilowatt-hour, m² = square meter, PV = photovoltaic.

Note: Numbers may not sum precisely because of rounding.

Source: Asian Development Bank estimates.

**Figure A1.1:** Average Monthly Solar Radiation in Kurendhoo Island

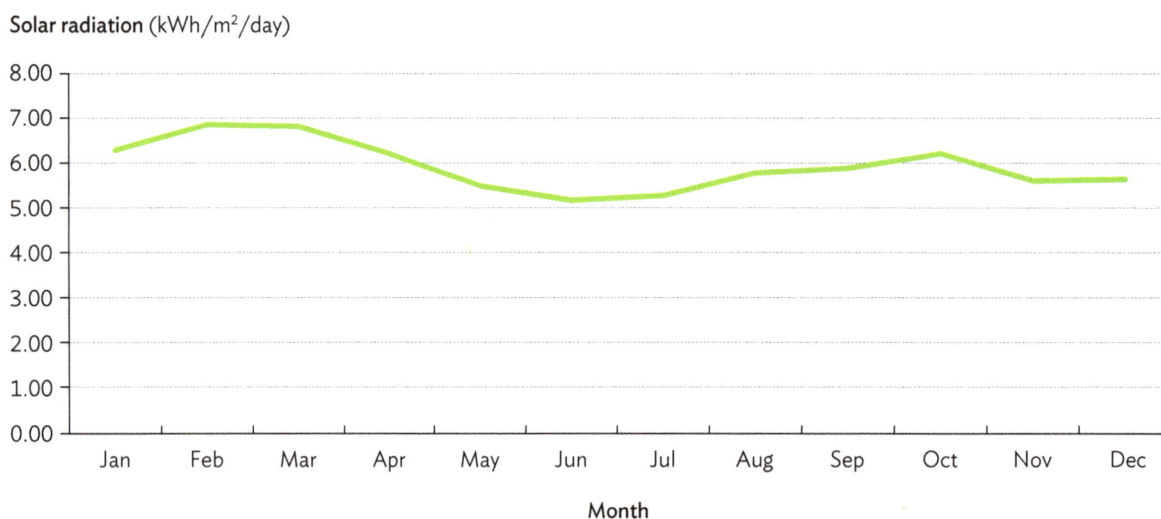

kWh = kilowatt-hour, m² = square meter.

Source: Asian Development Bank estimates.

**Figure A1.2:** Monthly Electricity Generation Profile of the 107-Kilowatt-Peak Solar Photovoltaic System in Kurendhoo Island

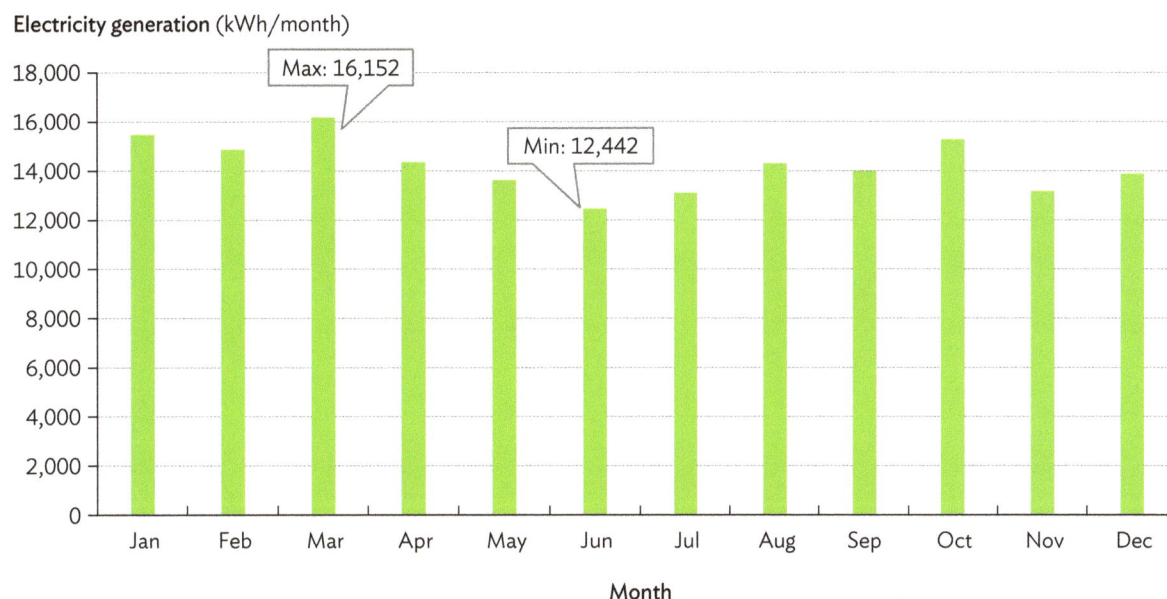

Electricity generation (kWh/month)

Max: 16,152

Min: 12,442

[Bar chart showing monthly electricity generation from Jan to Dec, with values ranging between approximately 12,442 and 16,152 kWh/month]

Month

kWh = kilowatt-hour.

Source: Asian Development Bank estimates.

Table A1.3 presents the results of the cost–benefit analysis of the solar and battery energy storage systems in Kurendhoo Island.

**Table A1.3:** Costs, Net Benefit, and Economic Internal Rate of Return of the Solar and Battery Energy Storage System Investment in Kurendhoo Island

| Year | Cost (Rf) | | Benefit (Rf) | Net Benefit (Rf) |
|---|---|---|---|---|
| | Capital | O&M | Annual Savings | |
| 0 | 3,902,237 | | | –3,902,237 |
| 1 | | 39,022 | 969,152 | 930,130 |
| 2 | | 39,022 | 959,460 | 920,438 |
| 3 | | 39,022 | 949,866 | 910,844 |
| 4 | | 39,022 | 940,367 | 901,345 |
| 5 | | 39,022 | 930,964 | 891,941 |
| 6 | | 39,022 | 921,654 | 882,632 |
| 7 | | 39,022 | 912,437 | 873,415 |
| 8 | | 39,022 | 903,313 | 864,291 |

*continued on next page*

| Year | Cost (Rf) | | Benefit (Rf) | Net Benefit (Rf) |
|---|---|---|---|---|
| | Capital | O&M | Annual Savings | |
| 9 | | 39,022 | 894,280 | 855,257 |
| 10 | | 39,022 | 885,337 | 846,315 |
| 11 | | 39,022 | 876,484 | 837,461 |
| 12 | | 39,022 | 867,719 | 828,696 |
| 13 | 390,224 | 39,022 | 859,042 | 429,796 |
| 14 | | 39,022 | 850,451 | 811,429 |
| 15 | | 39,022 | 841,947 | 802,924 |
| 16 | | 39,022 | 833,527 | 794,505 |
| 17 | | 39,022 | 825,192 | 786,170 |
| 18 | | 39,022 | 816,940 | 777,918 |
| 19 | | 39,022 | 808,771 | 769,748 |
| 20 | | 39,022 | 800,683 | 761,661 |
| 21 | | 39,022 | 792,676 | 753,654 |
| 22 | | 39,022 | 784,749 | 745,727 |
| 23 | | 39,022 | 776,902 | 737,880 |
| 24 | | 39,022 | 769,133 | 730,111 |
| 25 | | 39,022 | 761,442 | 722,419 |
| | | | **EIRR** | **23%** |

EIRR = economic internal rate of return, O&M = operation and maintenance, Rf = rufiyaa.
Note: Numbers may not sum precisely because of rounding.
Source: Asian Development Bank estimates.

# APPENDIX 2

## SOLAR PHOTOVOLTAIC POWER OUTPUT IN ADDU CITY

Table A2.1 provides the detailed specifications of the upgraded hybrid power system implemented by the Preparing Outer Islands for Sustainable Energy Development (POISED) project in Addu City.

**Table A2.1: Specifications for Solar Photovoltaic Generation in Addu City**

| Parameter | Hithadhoo, Maldives |
|---|---|
| Latitude (degree N) | 0.7 |
| Longitude (degree E) | 73.2 |
| Elevation (meter) | 2 |
| DC System Size (kilowatt) | 1,600 |
| Module type | Standard |
| Array type | Fixed (open rack) |
| Array tilt (degree) | 8 |
| Array Azimuth (degree) | 180 |
| System losses | 14.1 |
| Inverter efficiency | 96.0 |
| DC to AC ratio | 1.2 |

AC = alternating current, DC = direct current.
Source: Fenaka Corporation Ltd.

The estimated electricity generation of the 1,600 kilowatt-peak (kWp) solar photovoltaic system in Addu City is presented in Table A2.2. Figure A2.1 shows the average monthly solar radiation in Addu City, and Figure A2.2 shows the variation in the solar power generation profile for the 1,600 kWp system.

It is estimated that the average annual solar radiation in Addu City is 5.48 kilowatt-hour per square meter per day (kWh/m²/day). The typical variation is between 4.69 kWh/m²/day and 6.30 kWh/m²/day. The maximum (219,143 kWh/month) electricity generation can be expected in March and the minimum (173,054 kWh/month) in June.

**Table A2.2:** Estimated Monthly Electricity Generation of the 1,600-Kilowatt-Peak Solar Photovoltaic System in Addu City

| Month | Solar Radiation (kWh/m²/day) | AC System Output (kWh) |
|---|---|---|
| January | 5.73 | 211,903 |
| February | 6.30 | 206,920 |
| March | 6.02 | 219,143 |
| April | 5.57 | 198,556 |
| May | 4.96 | 186,119 |
| June | 4.82 | 173,054 |
| July | 4.69 | 174,013 |
| August | 4.96 | 186,352 |
| September | 5.52 | 198,826 |
| October | 5.64 | 210,004 |
| November | 5.91 | 211,884 |
| December | 5.69 | 210,462 |
| **Average** | **5.48** | **198,936** |

AC = alternating current, kWh = kilowatt-hour, m² = square meter.

Source: Asian Development Bank estimates.

**Figure A2.1:** Average Monthly Solar Radiation in Addu City

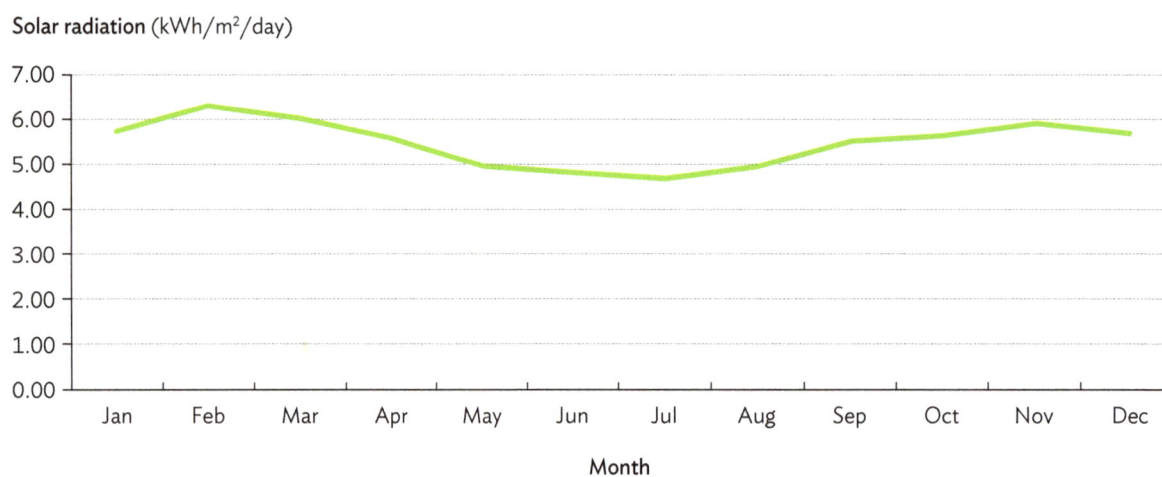

kWh = kilowatt-hour, m² = square meter.

Source: Asian Development Bank estimates.

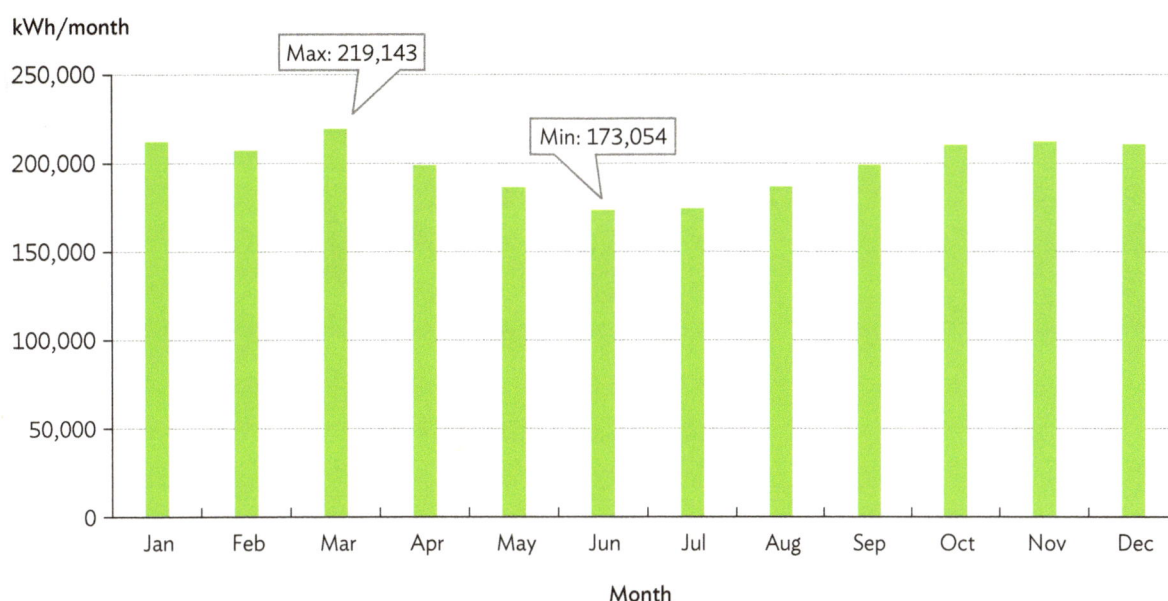

**Figure A2.2:** Monthly Electricity Generation Profile of the 1,600-Kilowatt-Peak Solar Photovoltaic System in Addu City

kWh/month

Max: 219,143

Min: 173,054

Month

kWh = kilowatt-hour.
Source: Asian Development Bank estimates.

Table A2.3 presents the results of the economic cost–benefit analysis of the advanced battery energy storage system and energy management system in Addu City.

**Table A2.3:** Economic Cost–Benefit Analysis of the Upgraded Hybrid Power System in Addu City

| Year | Cost (Rf) | | | | Benefit (Rf) | Net Benefit (Rf) | |
| | Capital | | O&M | | | | |
| | 3C | 1C | 3C | 1C | Fuel Savings | 3C | 1C |
|---|---|---|---|---|---|---|---|
| | 50,000,000 | 55,000,000 | | | | −50,000,000 | −55,000,000 |
| 1 | | | 250,000 | 275,000 | 12,268,792 | 12,018,792 | 11,993,792 |
| 2 | | | 250,000 | 275,000 | 12,146,104 | 11,896,104 | 11,871,104 |
| 3 | | | 250,000 | 275,000 | 12,024,643 | 11,774,643 | 11,749,643 |
| 4 | | | 250,000 | 275,000 | 11,904,397 | 11,654,397 | 11,629,397 |
| 5 | | | 250,000 | 275,000 | 11,785,353 | 11,535,353 | 11,510,353 |
| 6 | | | 250,000 | 275,000 | 11,667,499 | 11,417,499 | 11,392,499 |
| 7 | | | 250,000 | 275,000 | 11,550,824 | 11,300,824 | 11,275,824 |
| 8 | | 5,793,750 | 250,000 | 275,000 | 11,435,316 | 11,185,316 | 5,366,566 |

*continued on next page*

| | Cost (Rf) | | | | Benefit (Rf) | Net Benefit (Rf) | |
|---|---|---|---|---|---|---|---|
| | Capital | | O&M | | | | |
| Year | 3C | 1C | 3C | 1C | Fuel Savings | 3C | 1C |
| 9 | | | 250,000 | 275,000 | 11,320,963 | 11,070,963 | 11,045,963 |
| 10 | | | 250,000 | 275,000 | 11,207,753 | 10,957,753 | 10,932,753 |
| 11 | | | 250,000 | 275,000 | 11,095,676 | 10,845,676 | 10,820,676 |
| 12 | | | 250,000 | 275,000 | 10,984,719 | 10,734,719 | 10,709,719 |
| 13 | 5,000,000 | 5,500,000 | 250,000 | 275,000 | 10,874,872 | 5,624,872 | 5,099,872 |
| 14 | | | 250,000 | 275,000 | 10,766,123 | 10,516,123 | 10,491,123 |
| 15 | 5,021,250 | | 250,000 | 275,000 | 10,658,462 | 5,387,212 | 10,383,462 |
| 16 | | 5,793,750 | 250,000 | 275,000 | 10,551,877 | 10,301,877 | 4,483,127 |
| 17 | | | 250,000 | 275,000 | 10,446,358 | 10,196,358 | 10,171,358 |
| 18 | | | 250,000 | 275,000 | 10,341,895 | 10,091,895 | 10,066,895 |
| 19 | | | 250,000 | 275,000 | 10,238,476 | 9,988,476 | 9,963,476 |
| 20 | | | 250,000 | 275,000 | 10,136,091 | 9,886,091 | 9,861,091 |
| 21 | | | 250,000 | 275,000 | 10,034,730 | 9,784,730 | 9,759,730 |
| 22 | | | 250,000 | 275,000 | 9,934,383 | 9,684,383 | 9,659,383 |
| 23 | | | 250,000 | 275,000 | 9,835,039 | 9,585,039 | 9,560,039 |
| 24 | | 5,793,750 | 250,000 | 275,000 | 9,736,689 | 9,486,689 | 3,667,939 |
| 25 | | | 250,000 | 275,000 | 9,639,322 | 9,389,322 | 9,364,322 |
| | | | | | **EIRR** | **23%** | **20%** |

C = cycle, EIRR = economic internal rate of return, O&M = operation and maintenance, Rf = rufiyaa.
Source: Asian Development Bank estimates.

# BIBLIOGRAPHY

Asian Development Bank (ADB). 2008. *Strategy 2020: Working for an Asia and Pacific Free of Poverty.*

———. 2012. *Country Operations Business Plan: Maldives, 2013–2015.*

———. 2020. *A Brighter Future for Maldives Powered by Renewables: Road Map for the Energy Sector 2020–2030.*

Cole, W., A.W. Frazier, and C. Augustine. 2021. *Cost Projections for Utility-Scale Battery Storage: 2021 Update.* National Renewable Energy Laboratory.

Fenaka Corporation Ltd. 2018. *Central Power Station, Addu City, Maldives – Power System Evaluation Report.*

Government of Maldives. 2019. Strategic Action Plan 2019–2023.

Government of Maldives, Ministry of Environment. 2019. *A Shift Towards Clean Energy in Maldives.*

———. 2019. *Maldives First Biennial Update Report to the United Nations Framework Convention on Climate Change.*

———. 2020. *Update of Nationally Determined Contributions of Maldives.*

Government of Maldives, Ministry of Environment and Energy. 2014. *Proposed Grant and Administration of Grant—Republic of Maldives: Preparing Outer Islands for Sustainable Energy Development.*

———. 2015. *Maldives Climate Change Policy Framework.*

Government of Maldives, Ministry of Housing and Environment. 2010. *Maldives National Energy Policy and Strategy.*

Government of Maldives, Ministry of Planning and National Development. 2007. *Seventh National Development Plan 2006–2010: Creating New Opportunities.*

Government of Maldives, Utility Regulatory Authority. 2019. *Tariff Revision 2019.*

Kersey, J., M. Sprengel, G. Babbitt, and T. Johnson. 2017. *Hybrid Power Generation for Improved Fuel Efficiency and Performance.* Conference Proceedings and Presentations of Electrical Energy Storage Applications and Technologies.

Kolantharaj, J. et al. 2019. Advanced Battery Technology to Integrate Intermittent Renewables in the Maldives. *Development Asia.* 15 November.

Reşitoğlu, İ.A. et al. 2014. The Pollutant Emissions from Diesel-Engine Vehicles and Exhaust Aftertreatment Systems. *Clean Technologies and Environmental Policy.* 7. pp. 15–27.

Toshiba Energy Systems and Solutions Corporation. 2020. *EMS Operation Flow.* PowerPoint presentation.

United Nations Development Programme. 2007. *Overcoming Vulnerability to Rising Oil Prices: Options for Asia and the Pacific.*

World Bank. 2019. *Accelerating Renewable Energy Integration and Sustainable Energy.* Project Information Document.